mindsight

"In his new, graceful, wise, creative, utterly approachable book, *Mindsight*, Dr. Daniel Siegel integrates two of the most important discoveries of our time: the cutting-edge research into the brain functions relevant to understar  ery that our brains are plastic and c rly guided mental activity. Throug se histories, he shows us the princ nd ourselves and, more often than ls, brains, relationships, some life and even the course of some important mental illnesses."

Norman Doidge, author of *The Brain that Changes Itself*

"Dr. Siegel, an internationally acclaimed psychiatrist and psychotherapist, has provided us with an extraordinary rich and scholarly guide into the human mind and its capacity for understanding itself and the minds of others. Ranging over areas such as the importance of compassion and affection in our lives from the day we are born, and the transforming power of mindfulness, his immense knowledge and experience open up new horizons for scientific study and personal transformation. Written with personal anecdotes, gentleness, and wisdom, the ease and clarity of his writing is a dream. He never patronises us but always educates. This book is a great gift to those who seek to bring more wisdom and compassion to the world and digs deeply into the roots of our suffering. A treasure of a book."

Paul Gilbert, Professor of Clinical Psychology at the University of Derby and author of *The Compassionate Mind*

"An extraordinary book, integrating the knowledge from the latest brain science with ancient wisdom. Daniel Siegel offers both new insights and practices that can transform lives. A wonderful achievement."

Mark Williams, Professor of Clinical Psychology and Wellcome Principal Research Fellow, University of Oxford Department of Psychiatry

"Dr. Siegel's groundbreaking explorations of the brain provide a remarkable window into the physiological underpinnings of human behaviour. *Mindsight* is a literary MRI: a mind-blowing book that will change the way you think about the way you think."

Arianna Huffington, co-founder of *The Huffington Post*

"Drawing upon and explaining the intricate workings of the brain, *Mindsight* sets itself apart from other self-help books. Dr. Siegel helps the reader understand how we can change our dysfunctional habits of mind and become more flexible, adaptive, coherent, energized, and stable. He helps us see that we can rewire our own brains and become truly integrated, through personal understanding and, most important, through meaningful relationships with others. This is a must-read for anyone who wishes to have a happier, more productive life."

Eugene Beresin, M.D., Professor of Psychiatry, Harvard Medical School

"In *The Developing Mind*, Daniel Siegel brilliantly revealed how relationships sculpt and are sculpted by the brain. *Mindsight* is the perfect follow-up, a daring plan of action for a wiser and kinder life that's transformative yet easy to understand, and should appeal to specialists and laymen alike – in fact, to anyone who wants to enrich their life, their children's lives, and society."

Diane Ackerman, author of *A Natural History of the Senses*

"*Mindsight* is a rare book. Rooted in groundbreaking scientific research and searching professional practice, it is also a deeply compassionate and human account of what it is to be human. *Mindsight* has powerful lessons for doctors, for parents and educators, and for all of us who are trying to make sense of how we make sense of things."

Sir Ken Robinson, author of *The Element: A New View of Human Capacity*

"*Mindsight* is a remarkable exploration into the synergistic workings of the mind and the brain...It will be enormously useful for patients and their families, as well as mental health workers and the lay public at large."

Clarice Kestenbaum, Professor of Clinical Psychiatry at Columbia University

"In this brilliant and highly readable book, Dr. Siegel combines his prodigious knowledge of brain science, clinical psychology, and mindfulness with his immense capacity for original thinking to develop a new and useful concept: mindsight...His work will forever change the way we understand ourselves and our relationships."

Mary Pipher, author of *Reviving Ophelia* and *Seeking Peace*

"Right now, Dr. Siegel is creating a stir among therapists unmatched by any other in the field. *Mindsight* offers a fascinating synthesis of his innovative ideas about the implications of the new brain science for understanding relationships and the processes of human change."

Richard Simon, editor of *Psychotherapy Networker*

"An extraordinary and practical wedding of neuroscience and spiritual wisdom. Accessible and visionary, *Mindsight* is bound to be a classic."

Jack Kornfield, author of *The Wise Heart*

"This exciting book reveals the secrets of the mind that we have sought in Eastern and Western thought for two thousand years. How do we see the mind and learn to tame it for a happier and healthier life? Filled with engaging stories, *Mindsight* uses cutting-edge science and deep humanity to address the questions that we all have about the mystery in our skull."

Natalie Goldberg, author of *Old Friend from Far Away*

"*Mindsight* is a seminal piece on bringing neuroscience to everyday life, helping us to understand what can go awry in the mind so that, armed with that knowledge, we will be better able to change...Dr. Daniel Siegel's use of elaborate personal as well as patient stories makes us feel as though we are on a guided tour with a friendly group of fellow travellers."

John J. Ratey, author of *Spark: The Revolutionary New Science of Exercise and the Brain*

"Relevant, utterly compelling, and even life-altering."

Jon Kabat-Zinn, Ph.D., author of *Full Catastrophe Living*

"Integrating ancient contemplative practice with contemporary neuroscience and psychotherapy, Dan Siegel removes the veil on the mysteries of the interface between mind, brain, and relationships with novel and profound descriptions of how we become who we are, what makes things go wrong and how all of us can achieve optimal well being by using our capacity for attention to change the very structure and function of our brains...A book that reads like a mystery novel."

Harville Hendrix, Ph. D., author of *Getting The Love You Want: A Guide for Couples*

"[A] revolutionary book."

The Boston Globe

"An exciting exploration of how a troubled mind can right itself... Siegel's method isn't a quick fix and doesn't sugarcoat reality: The mindful traits of serenity, courage and wisdom involve accepting our place in the order of things. He challenges his patients to a life of tough work and convincingly suggests it will be well worth the effort."

Publishers Weekly

"Drawing on cutting-edge neurobiological research and Eastern meditation practices as well as studies conducted by his own, L.A.-based Mindsight Institute, Siegel presents a convincing case that mindsight's dual focus on mindfulness and empathy can literally rewire the brain and catalyze greater personal fulfillment. In 12 lucid yet scientifically grounded chapters, he provides the evidence for mindsight's powerful effect on human behavior, and then presents a guidebook for developing and applying mindsight in one's life. Unlike his earlier, more academic works, *Mindsight* is refreshingly accessible, offering solid practical advice while avoiding the naive optimism of many mainstream self-help books."

Booklist

"Siegel's clinical explanations make fascinating reading."

The Psychologist

mindsight

TRANSFORM YOUR BRAIN WITH THE NEW SCIENCE OF KINDNESS

DANIEL SIEGEL

ONEWORLD

A Oneworld Book

First published in Great Britain by Oneworld Publications 2010
This edition published 2011
Reprinted in 2013 (twice), 2014, 2016, 2018

This edition published by arrangement with Bantam Books, an
imprint of The Random House Publishing Group, a division of
Random House, Inc.

ISBN 978-1-85168-793-0
eISBN 978-1-85168-883-8

Printed and bound in Great Britain by Clays Ltd, St Ives plc

Oneworld Publications
10 Bloomsbury Street
London WC1B 3SR
England

To the two wondrous individuals who call me Dad

and

To my patients, past and present,
who have taught me so much about courage and transformation

Foreword

THE GREAT LEAPS FORWARD IN PSYCHOLOGY have come from original insights that suddenly clarify our experience from a fresh angle, revealing hidden patterns of connection. Freud's theory of the unconscious and Darwin's model of evolution continue to help us understand the findings from current research on human behavior and some of the mysteries of our daily lives. Daniel Siegel's theory of mindsight—the brain's capacity for both insight and empathy— offers a similar "Aha!" He makes sense for us out of the cluttered confusions of our sometimes maddening and messy emotions.

Our ability to know our own minds as well as to sense the inner world of others may be the singular human talent, the key to nurturing healthy minds and hearts. I've explored this terrain in my own work on emotional and social intelligence. Self-awareness and empathy are (along with self-mastery and social skills) domains of human ability essential for success in life. Excellence in these capacities helps people flourish in relationships, family life, and marriage, as well as in work and leadership.

Of these four key life skills, self-awareness lays the foundation for the rest. If we lack the capacity to monitor our emotions, for example, we will be poorly suited to manage or learn from them. Tuned out of a range of our own experience, we will find it all the harder to attune to that same range in others. Effective interactions depend on the smooth integration of self-awareness, mastery, and empathy. Or so I've argued. Dr. Siegel casts the discussion in a fresh light, putting these dynamics in terms of mindsight, and marshals compelling evidence for its crucial role in our lives.

A gifted and sensitive clinician, as well as a master synthesizer of research findings from neuroscience and child development,

Dr. Siegel gives us a map forward. Over the years he has continually broken new ground in his writing on the brain, psychotherapy, and child-rearing; his seminars for professionals are immensely popular.

The brain, he reminds us, is a social organ. Mindsight is the core concept in "interpersonal neurobiology," a field Dr. Siegel has pioneered. This two-person view of what goes on in the brain lets us understand how our daily interactions matter neurologically, shaping neural circuits. Every parent helps sculpt the growing brain of a child; the ingredients of a healthy mind include an attuned, empathetic parent—one with mindsight. Such parenting fosters this same crucial ability in a child.

Mindsight plays an integrative role in the triangle connecting relationships, mind, and brain. As energy and information flow among these elements of human experience, patterns emerge that shape all three (and the brain here includes its extensions via the nervous system throughout the body). This vision is holistic in the true sense of the word, inclusive of our whole being. With mindsight we can better know and manage this vital flow of being.

Dr. Siegel's biographical details are impressive. Harvard-trained and a clinical professor of psychiatry at UCLA and co-director of the Mindful Awareness Research Center there, he also founded and directs the Mindsight Institute. But far more impressive is his actual being, a mindful, attuned, and nurturing presence that is nourishing in itself. Dr. Siegel embodies what he teaches.

For professionals who want to delve into this new science, I recommend Dr. Siegel's 1999 text on interpersonal neurobiology, *The Developing Mind: Toward a Neurobiology of Interpersonal Experience*. For parents, his book with Mary Hartzell is invaluable: *Parenting from the Inside Out: How a Deeper Self-Understanding Can Help You Raise Children Who Thrive*. But for anyone who seeks a more rewarding life, the book you hold in your hands has compelling and practical answers.

DANIEL GOLEMAN

Contents

Introduction
Diving into the Sea Inside

WITHIN EACH OF US there is an internal mental world—what I have come to think of as the sea inside—that is a wonderfully rich place, filled with thoughts and feelings, memories and dreams, hopes and wishes. Of course it can also be a turbulent place, where we experience the dark side of all those wonderful feelings and thoughts—fears, sorrows, dreads, regrets, nightmares. When this inner sea seems to crash in on us, threatening to drag us down below to the dark depths, it can make us feel as if we are drowning. Who among us has not at one time or another felt overwhelmed by the sensations from within our own minds? Sometimes these feelings are just a passing thing—a bad day at work, a fight with someone we love, an attack of nerves about a test we have to take or a presentation we have to give, or just an inexplicable case of the blues for a day or two. But sometimes they seem to be something much more intractable, so much part of the very essence of who we are that it may not even occur to us that we can change them. This is where the skill that I have called "mindsight" comes in, for mindsight, once mastered, is a truly transformational tool. Mindsight has the potential to free us from patterns of mind that are getting in the way of living our lives to the fullest.

WHAT IS MINDSIGHT?

Mindsight is a kind of focused attention that allows us to see the internal workings of our own minds. It helps us to be aware of our mental processes without being swept away by them, enables us to get ourselves off the autopilot of ingrained behaviors and habitual responses, and moves us beyond the reactive emotional loops we all have a tendency to get trapped in. It lets us "name and tame" the emotions we are experiencing, rather than being overwhelmed by them. Consider the difference between saying "I am sad" and "I feel sad." Similar as those two statements may seem, there is actually a profound difference between them. "I am sad" is a kind of self-definition, and a very limiting one. "I feel sad" suggests the ability to recognize and acknowledge a feeling, without being consumed by it. The focusing skills that are part of mindsight make it possible to see what is inside, to accept it, and in the accepting to let it go, and, finally, to transform it.

You can also think of mindsight as a very special lens that gives us the capacity to perceive the mind with greater clarity than ever before. This lens is something that virtually everyone can develop, and once we have it we can dive deeply into the mental sea inside, exploring our own inner lives and those of others. A uniquely human ability, mindsight allows us to examine closely, in detail and in depth, the processes by which we think, feel, and behave. And it allows us to reshape and redirect our inner experiences so that we have more freedom of choice in our everyday actions, more power to create the future, to become the author of our own story. Another way to put it is that mindsight is the basic skill that underlies everything we mean when we speak of having social and emotional intelligence.

Interestingly enough, we now know from the findings of neuroscience that the mental and emotional changes we can create through cultivation of the skill of mindsight are transformational at the very physical level of the brain. By developing the ability to focus our attention on our internal world, we are picking up a "scalpel" we can use to resculpt our neural pathways, stimulating the growth of areas of the brain that are crucial to mental health. I will talk a lot about this in the chapters that follow because I believe that a basic understanding of how the brain works helps people see how much potential there is for change.

But change never just happens. It's something we have to work at. Though the ability to navigate the inner sea of our minds—to have mindsight—is our birthright, and some of us, for reasons that will become clear later, have a lot more of it than others, it does not come automatically, any more than being born with muscles makes us athletes. The scientific reality is that we need certain experiences to develop this essential human capacity. I like to say that parents and other caregivers offer us our first swimming lessons in that inner sea, and if we've been fortunate enough to have nurturing relationships early in life, we've developed the basics of mindsight on which we can build. But even if such early support was lacking, there are specific activities and experiences that can nurture mindsight throughout the lifespan. As you will see, mindsight is a form of expertise that can be honed in each of us, whatever our early history.

When I first began to explore the nature of the mind professionally, there was no term in our everyday language that captured the way we perceive our thoughts, feelings, sensations, memories, beliefs, attitudes, hopes, dreams, and fantasies. Of course, these activities of the mind fill our day-to-day lives—we don't need to learn a skill in order to experience them. But how do we actually develop the ability to perceive a thought—not just have one—and to know it as an activity of our minds so that we are not taken over by it? How can we be receptive to the mind's riches and not just reactive to its reflexes? How can we direct our thoughts and feelings rather than be driven by them? And how can we know the minds of others, so that we truly understand "where they are coming from" and can respond more effectively and compassionately? When I was a young psychiatrist, there weren't many readily accessible scientific or even clinical terms to describe the whole of this ability. To be able to help my patients, I coined the term *mindsight* so that together we could discuss this important ability that allows us to see and shape the inner workings of our own minds.

Our first five senses allow us to perceive the outside world—to hear a bird's song or a snake's warning rattle, to make our way down a busy street or smell the warming earth of spring. What has been called our sixth sense allows us to perceive our internal bodily states—the quickly beating heart that signals fear or excitement, the sensation of butterflies in our stomach, the pain that demands our attention. Mindsight, our ability to look within and perceive the mind, to reflect on our experience, is every bit as essential to our well-being. Mindsight is our seventh sense.

As I hope to show you in this book, this essential skill can help us build social and emotional brainpower, move our lives from disorder to well-being, and create satisfying relationships filled with connection and compassion. Business and government leaders have told me that understanding how the mind functions in groups has helped them be more effective and enabled their organizations to become more productive. Clinicians in medicine and mental health have said that mindsight has changed the way they approach their patients, and that putting the mind at the heart of their healing work has helped them create novel and useful interventions. Teachers introduced to mindsight have learned to "teach with the brain in mind" and are reaching and teaching their students in deeper and more lasting ways.

In our individual lives, mindsight offers us the opportunity to explore the subjective essence of who we are, to create a life of deeper meaning with a richer and more understandable internal world. With mindsight we are better able to balance our emotions, achieving an internal equilibrium that enables us to cope with the small and large stresses of our lives. Through our ability to focus attention, mindsight also helps the body and brain achieve homeostasis—the internal balance, coordination, and adaptiveness that forms the core of health. Finally, mindsight can improve our relationships with our friends, colleagues, spouses, and children—and even the relationship we have with our own selves.

A NEW APPROACH TO WELL-BEING

Everything that follows rests on three fundamental principles. The first is that mindsight can be cultivated through very practical steps. This means that creating well-being—in our mental life, in our close relationships, and even in our bodies—is a learnable skill. Each chapter of this book explores these skills, from basic to advanced, for navigating the sea inside.

Second, as mentioned above, when we develop the skill of mindsight, we actually change the physical structure of the brain. Developing the lens that enables us to see the mind more clearly stimulates the brain to grow important new connections. This revelation is based on one of the most exciting scientific discoveries of the last twenty years: How we focus our attention shapes the structure of the brain. Neuroscience supports the idea that developing the reflective skills of mindsight activates the very circuits that create resilience and well-being and that underlie empathy and compassion as well. Neuroscience has also definitively shown that we can grow these new connections throughout our lives, not just in childhood. The short Minding the Brain sections interspersed throughout part 1 are a traveler's guide to this new territory.

The third principle is at the heart of my work as a psychotherapist, educator, and scientist. Well-being emerges when we create connections in our lives—when we learn to use mindsight to help the brain achieve and maintain *integration*, a process by which separate elements are linked together into a working whole. I know this may sound both unfamiliar and abstract at first, but I hope you'll soon find that it is a natural and useful way of thinking about our lives. For example, integration is at the heart of how we connect to one another in healthy ways, honoring one another's differences while keeping our lines of communication wide open. Linking separate entities to one another—integration—is also important for releasing the creativity that emerges when the left and right sides of the brain are functioning together.

Integration enables us to be flexible and free; the lack of such connections promotes a life that is either rigid or chaotic, stuck and dull on the one hand or explosive and unpredictable on the other. With the connecting freedom of integration comes a sense of vitality and the ease of well-being. Without integration we can become imprisoned in behavioral ruts—anxiety and depression, greed, obsession, and addiction.

By acquiring mindsight skills, we can alter the way the mind functions and move our lives towards integration, away from these extremes of chaos or rigidity. With mindsight we are able to focus our mind in ways that literally integrate the brain and move it towards resilience and health.

MINDSIGHT MISUNDERSTOOD

It's wonderful to receive an email from an audience member or patient who says, "My whole view of reality has changed." But not everyone new to mindsight gets it right away. Some people are concerned that it's just another way to become more self-absorbed—a form of navel-gazing, of becoming preoccupied with "reflection" instead of living fully. Perhaps you've also read some of the recent research (or the ancient wisdom) that tells us that happiness depends on "getting out of yourself." Does mindsight turn us away from this greater good? While it is true that being self-obsessed decreases happiness, mindsight actually frees you to become less self-absorbed, not more. When we are not taken over by our thoughts and feelings, we can become clearer in our own internal world as well as more receptive to the inner world of another. Scientific studies support this idea, revealing that individuals with more mindsight skills show more interest and empathy towards others. Research has also clearly shown that mindsight supports not only internal and interpersonal well-being but also greater effectiveness and achievement in school and work.

Another quite poignant concern about mindsight came up one day when I was talking with a group of teachers. "How can you ask us to have children reflect on their own minds?" one teacher said to me. "Isn't that opening a Pandora's box?" Recall that when Pandora's box was opened, all the troubles of humanity flew out. Is this how we imagine our inner lives or the inner lives of our children? In my own experience, a great transformation begins when we look at our minds with curiosity and respect rather than fear and avoidance. Inviting our thoughts and feelings into awareness allows us to learn from them rather than be driven by them. We can calm them without ignoring them; we can hear their wisdom without being terrified by their screaming voices. And as you will see in some of the stories in this book, even surprisingly young children can develop the ability to pause and make choices about how to act when they are more aware of their impulses.

HOW DO WE CULTIVATE MINDSIGHT?

Mindsight is not an all-or-nothing ability, something you either have or don't have. As a form of expertise, mindsight can be developed when we put in effort, time, and practice.

Most people come into the world with the brain potential to develop mindsight, but the neural circuits that underlie it need experiences to develop properly. For some—such as those with autism and related neurological conditions—the neural circuits of mindsight may not develop well even with the best caregiving. In most children, however, the ability to see the mind develops through everyday interactions with others, especially through attentive communication with parents and caregivers. When adults are in tune with a child, when they reflect back to the child an accurate picture of his internal world, he comes to sense his own mind with clarity. This is the foundation of mindsight. Neuroscientists are now identifying the circuits of the brain that participate in this intimate dance and exploring how a caregiver's attunement to the child's internal world stimulates the development of those neural circuits.

If parents are unresponsive, distant, or confusing in their responses, however, their lack of attunement means that they cannot reflect back to the child an accurate picture of the child's inner world. In this case, research suggests, the child's mindsight lens may become cloudy or distorted. The child may then be able to see only part of the sea inside, or see it dimly. Or the child may develop a lens that sees well but is fragile, easily disrupted by stress and intense emotions.

The good news is that whatever our early history, it is never too late to stimulate the growth of the neural fibers that enable mindsight to flourish. You'll soon meet a ninety-two-year-old man who was able to overcome a painful and twisted childhood to emerge a mindsight expert. Here we see living evidence for another exciting discovery of modern neuroscience: that the brain never stops growing in response to experience. And this is true for people with happy childhoods, too. Even if we had positive relationships with our caregivers and parents early on—and even if we write books on the subject—we can continue as long as we live to keep developing our vital seventh sense and promoting the connections and integration that are at the heart of well-being.

We'll begin our journey in part 1 by exploring situations in which the vital skills of mindsight are absent. These stories reveal how seeing the mind clearly and being able to alter how it functions are essential elements in the path towards well-being. Part 1 is the more theoretical section of the book, where I explain the basic concepts, give readers an introduction to brain science, and offer working definitions of the mind and mental health. Since I know that my readers will come from a wide variety of backgrounds and interests, I realize that some of you may want to skim or even skip much of that material in order to move directly to part 2. In part 2, we'll dive deeply into stories from my practice that illustrate the steps involved in developing the skills of mindsight. This is the section of the book in which I share the knowledge and practical skills that will help people understand how to shape their own minds towards health. At the very end of the book is an appendix outlining the fundamental concepts and a set of endnotes with the scientific resources supporting these ideas.

Our exploration of mindsight begins with the story of a family that changed my own life and my entire approach to psychotherapy. Looking for ways to help them inspired me to search for new answers to some painful questions about what happens when mindsight is lost. It also led to my search for the techniques that can enable us to reclaim and recreate mindsight in ourselves, our children, and our communities. I hope you'll join me on this journey into the inner sea. Within those depths awaits a vast world of possibility.

PART I

· · · · · · · ·

The Path to Well-Being:
Mindsight Illuminated

1

A Broken Brain, a Lost Soul

The Triangle of Well-Being

BARBARA'S FAMILY MIGHT NEVER HAVE COME for therapy if seven-year-old Leanne hadn't stopped talking in school. Leanne was Barbara's middle child, between Amy, who was fourteen, and Tommy, who was three. They had all taken it hard when their mother was in a near-fatal car accident. But it wasn't until Barbara returned home from the hospital and rehabilitation center that Leanne became "selectively mute." Now she refused to speak with anyone outside the family—including me.

In our first weekly therapy sessions, we spent our time in silence, playing some games, doing pantomimes with puppets, drawing, and just being together. Leanne wore her dark hair in a single jumbled ponytail, and her sad brown eyes would quickly dart away whenever I looked directly at her. Our sessions felt stuck, her sadness unchanging, the games we played repetitive. But then one day when we were playing catch, the ball rolled to the side of the couch and Leanne discovered my video player and screen. She said nothing, but the sudden alertness of her expression told me her mind had clicked on to something.

The following week Leanne brought in a videotape, walked over to the video machine, and put it into the slot. I turned on the player and her smile lit up the room as we watched her mother gently lift a younger Leanne up into the air, again and again, and then pull her into a huge, enfolding hug, the two of them shaking with laughter from head to toe. Leanne's father, Ben, had captured on film the dance of communication between parent and child that is the hallmark of love: We connect with each other through a give-and-take of signals that link us from the inside out. This is the joy-filled way in which we come to share each other's minds.

Next the pair swirled around on the lawn, kicking the brilliant yellow and burnt-orange leaves of autumn. The mother-daughter duet approached the camera, pursed lips blowing kisses into the lens, and then burst out in laughter. Five-year-old Leanne shouted, "Happy birthday, Daddy!" at the top of her lungs, and you could see the camera shake as her father laughed along with the ladies in his life. In the background Leanne's baby brother, Tommy, was napping in his pushchair, snuggled under a blanket and surrounded by soft toys. Leanne's older sister, Amy, was off to the side engrossed in a book.

"That's how my mum used to be when we lived in Boston," Leanne said suddenly, the smile dropping from her face. It was the first time she had spoken directly to me, but it felt more like I was overhearing her talk to herself. Why had Leanne stopped talking?

It had been two years since that birthday celebration, eighteen months since the family moved to Los Angeles, and twelve months since Barbara suffered a severe brain injury in her accident—a head-on collision. Barbara had not been wearing her seat belt that evening as she drove their old Mustang to the local store to get some milk for the kids. When the drunk driver plowed into her, her forehead was forced into the steering wheel. She had been in a coma for weeks following the accident.

After she came out of the coma, Barbara had changed in dramatic ways. On the videotape I saw the warm, connected, and caring person that Barbara had been. But now, Ben told me, she "was just not the same Barbara anymore." Her physical body had come home, but Barbara herself, as they had known her, was gone.

During Leanne's next visit I asked for some time alone with her parents. It was clear that what had been a close relationship between Barbara and Ben was now profoundly stressed and distant. Ben was patient and kind with Barbara and seemed to care for her deeply, but I could sense his despair. Barbara just stared off as we talked, made little eye contact with either of us, and seemed to lack interest in the conversation. The damage to her forehead had been repaired by plastic surgery, and although she had been left with motor skills that were somewhat slow and clumsy, she actually looked quite similar, in outward appearance, to her image on the videotape. Yet something huge had changed inside.

Wondering how she experienced her new way of being, I asked Barbara what she thought the difference was. I will never forget her reply: "Well, I guess if you had to put it into words, I suppose I'd say that I've lost my soul."

Ben and I sat there, stunned. After a while, I gathered myself enough to ask Barbara what losing her soul felt like.

"I don't know if I can say any more than that," she said flatly. "It feels fine, I guess. No different. I mean, just the way things are. Just empty. Things are fine."

We moved on to practical issues about care for the children, and the session ended.

A DAMAGED BRAIN

It wasn't clear yet how much Barbara could or would recover. Given that only a year had passed since the accident, much neural repair was still possible. After an injury, the brain can regain some of its function and even grow new neurons and create new neural connections, but with extensive damage it may be difficult to retrieve the complex abilities and personality traits that were dependent on the now destroyed neural structures.

Neuroplasticity is the term used to describe this capacity for creating new neural connections and growing new neurons in response to experience. Neuroplasticity is not just available to us in youth: We now know that it can occur throughout the lifespan. Efforts at rehabilitation for Barbara would need to harness the power of neuroplasticity to grow the new connections that might be able to reestablish old mental functions. But we'd have to wait awhile for the healing effects of time and rehabilitation to see how much neurological recovery would be possible.

My immediate task was to help Leanne and her family understand how someone could be alive and look the same yet have become so radically different in the way her mind functioned. Ben had told me earlier that he did not know how to help the children deal with how

Barbara had changed; he said that he could barely understand it himself. He was on double duty, working, managing the kids' schedules, and making up for what Barbara could no longer do. This was a mother who had delighted in making homemade Halloween costumes and Valentine's Day cupcakes. Now she spent most of the day watching TV or wandering around the neighborhood. She could walk to the grocery store, but even with a list she would often come home empty-handed. Amy and Leanne didn't mind so much that she cooked a few simple meals over and over again. But they were upset when she forgot their special requests, things they'd told her they liked or needed for school. It was as if nothing they said to her really registered.

As our therapy sessions continued, Barbara usually sat quietly, even when she was alone with me, although her speech was intact. Occasionally she'd suddenly become agitated at an innocent comment from Ben, or yell if Tommy fidgeted or Leanne twirled her ponytail around her finger. She might even erupt after a silence, as if some internal process was driving her. But most of the time her expression seemed frozen, more like emptiness than depression, more vacuous than sad. She seemed aloof and unconcerned, and I noticed that she never spontaneously touched either her husband or her children. Once, when three-year-old Tommy climbed onto her lap, she briefly put her hand on his leg as if repeating some earlier pattern of behavior, but the warmth had gone out of the gesture.

When I saw the children without their mother, they let me know how they felt. "She just doesn't care about us like she used to," Leanne said. "And she doesn't ever ask us anything about ourselves," Amy added with sadness and irritation. "She's just plain selfish. She doesn't want to talk to anyone anymore." Tommy remained silent. He sat close to his father with a drawn look on his face.

Loss of someone we love cannot be adequately expressed with words. Grappling with loss, struggling with disconnection and despair, fills us with a sense of anguish and actual pain. Indeed, the parts of our brain that process physical pain overlap with the neural centers that record social ruptures and rejection. Loss rips us apart.

Grief allows you to let go of something you've lost only when you begin to accept what you now have in its place. As our mind clings to the familiar, to our established expectations, we can become trapped in feelings of disappointment, confusion, and anger that create our own internal worlds of suffering. But what were Ben and the kids actually letting go of? Could Barbara regain her connected way of being? How could the family learn to live with a person whose body was still alive, but whose personality and "soul"—at least as they had known her—were gone?

"YOU-MAPS" AND "ME-MAPS"

Nothing in my formal training—whether in medical school, pediatrics, or psychiatry—had prepared me for the situation I now faced in my treatment room. I'd had courses on brain anatomy and on brain and behavior, but when I was seeing Barbara's family, in the early 1990s, relatively little was known about how to bring our knowledge of such subjects into the clinical practice of psychotherapy. Looking for some way to explain Barbara to her family, I trekked to the medical library and reviewed the recent clinical and scientific literature that dealt with the regions of the brain damaged by her accident.

Scans of Barbara's brain revealed substantial trauma to the area just behind her forehead; the lesions followed the upper curve of the steering wheel. This area, I discovered, facilitates very important functions of our personality. It also links widely separated brain regions to one another—it is a profoundly integrative region of the brain.

The area behind the forehead is a part of the frontal lobe of the cerebral cortex, the outermost section of the brain. The frontal lobe is associated with most of our complex thinking and planning. Activity in this part of the brain fires neurons in patterns that enable us to form neural representations—"maps" of various aspects of our world. The maps resulting from these clusters of neuronal activity serve to create an image in our minds. For example, when we take in the light reflected from a bird sitting in a tree, our eyes send signals back into our brain, and the neurons there fire in certain patterns that permit us to have the visual picture of the bird.

Somehow, in ways still to be discovered, the physical property of neurons firing helps to create our subjective experience—the thoughts, feelings, and associations evoked by seeing that bird, for example. The sight of the bird may cause us to feel certain emotions, to hear or remember its song, and even to associate that song with ideas such as nature, hope, freedom, and peace. The more abstract and symbolic the representation, the higher in the nervous system it is created, and the more forward in the cortex.

The prefrontal cortex—the most damaged part of the frontal lobe of Barbara's brain—makes complex representations that permit us to create concepts in the present, think of experiences in the past, and plan and make images about the future. The prefrontal cortex is also responsible for the neural representations that enable us to make images of the mind itself. I call these representations of our mental world "mindsight maps." And I have identified several kinds of mindsight maps made by our brains.

The brain makes what I call a "me-map" that gives us insight into ourselves, and a "you-map" for insight into others. We also seem to create "we-maps," representations of our relationships. Without such maps, we are unable to perceive the mind within ourselves or others. Without a me-map, for example, we can become swept up in our thoughts or flooded by our feelings. Without a you-map, we see only others' behaviors, the physical aspect of reality, without sensing the subjective core, the inner mental sea of others. It is the you-map that permits us to have empathy. In essence, the injury to Barbara's brain had created a world without mindsight. She had feelings and thoughts, but she could not represent them to herself as activities of her mind. Even when she said she'd "lost her soul," her statement had a bland, factual quality, more like a scientific observation than a deeply felt expression of personal identity. (I was puzzled by that disconnect between observation and emotion until I learned from later studies that the parts of our brain that create maps of the mind are distinct from those that enable us to observe and comment on self-traits such as shyness or anxiety—or, in Barbara's case, the lack of a quality she called "soul.")

In the years since I took Barbara's brain scans to the library, much more has been discovered about the interlinked functions of the prefrontal cortex. For example, the side of this region is crucial for how we pay attention; it enables us to put things in the "front of our mind" and hold them in awareness. The middle portion of the prefrontal area, the part damaged in Barbara, coordinates an astonishing number of essential skills, including regulating the body, attuning to others, balancing emotions, being flexible in our responses, soothing fear, and creating empathy, insight, moral awareness, and intuition. These were the skills Barbara was no longer able to recruit in her interactions with her family.

I will be referring to—and expanding on—this list of nine middle prefrontal functions throughout our discussion of mindsight. But even at first glance, you can see that these functions are essential ingredients for well-being, ranging from bodily processes such as regulating our hearts to social functions such as empathy and moral reasoning.

After Barbara emerged from her coma, her impairments had seemed to settle into a new personality. Some of her habits, such as what she liked to eat and how she brushed her teeth, remained the same. There was nothing significantly changed in how her brain mapped out these basic behavioral functions. But the ways in which she thought, felt, behaved, and interacted with others were profoundly altered. This affected every detail of daily life—right down to Leanne's crooked ponytail. Barbara still had the behavioral moves necessary to fix her daughter's hair, but she no longer cared enough to get it right.

Above all, Barbara seemed to have lost the very map-making ability that would enable her to honor the reality and importance of her own or others' subjective inner lives. Her mindsight maps were no longer forming amid the now-jumbled middle prefrontal circuitry upon which they depended for their creation. This middle prefrontal trauma had also disrupted the communication between Barbara and her family—she could neither send nor receive the connecting signals enabling her to join minds with the people she had loved most.

Ben summed up the change: "She is gone. The person we live with is just not Barbara."

A TRIANGLE OF WELL-BEING: MIND, BRAIN, AND RELATIONSHIPS

The videotape of Ben's birthday had revealed a vibrant dance of communication between Barbara and Leanne. But now there was no dance, no music keeping the rhythm of two minds flowing into a sense of a "we." Such joining happens when we attune to the internal shifts in another person, as they attune to us, and our two worlds become linked as one. Through facial expressions and tones of voice, gestures and postures—some so fleeting they can be captured only on a slowed-down recording—we come to "resonate" with one another. The whole we create together is truly larger than our individual identities. We feel this resonance as a palpable sense of connection and aliveness. This is what happens when our minds meet.

A patient of mine once described this vital connection as "feeling felt" by another person: We sense that our internal world is shared, that our mind is *inside* the other. But Leanne no longer "felt felt" by her mum.

The way Barbara behaved with her family reminded me of a classic research tool used to study infant-parent communication and attachment. Called the "still-face" experiment, it is painful both to participate in and to watch.

A mother is asked to sit with her four-month-old infant facing her and when signaled, to stop interacting with her child. This "still" phase in which no verbal or nonverbal signals are to be shared with the child is profoundly distressing. For up to three minutes, the child attempts to engage the now-nonresponsive parent in a bid for connection. At first the child usually amps up her signals, increasing smiles, coos, eye contact. But after a period of continuing nonresponse, she becomes agitated and distressed, her organized bids for connection melting into signs of anguish and outrage. She may then attempt to soothe herself by placing her hand in her mouth or pulling at her clothes. Sometimes researchers or parents call off the experiment at this time, but sometimes it goes on until the infant withdraws, giving up in a kind of despondent collapse that looks like melancholic depression. These stages of protest, self-soothing, and despair reveal how much the child depends upon the attuned responses of a parent to keep her own internal world in equilibrium.

We come into the world wired to make connections with one another, and the subsequent neural shaping of our brain, the very foundation of our sense of self, is built upon these intimate exchanges between the infant and her caregivers. In the early years this interpersonal regulation is essential for survival, but throughout our lives we continue to need such connections for a sense of vitality and well-being.

Leanne once had an attuned mother, and Barbara's earlier presence in her life was now literally embedded in the structure of Leanne's mindsight-map-making brain. But Barbara could no longer map Leanne's mind, she could not feel her children within her, and she could not make them "feel felt." Her lack of interest in them, her apparent indifference to their feelings and needs, her withdrawal of what they experienced as love, was the outer sign of this inner tragedy.

Treating Barbara's family made it clear to me that mind, brain, and relationships are not separate elements of life—they are irreducible aspects of one interconnected triangle of well-being. At seven years of age, Leanne had responded to her mother's withdrawal by going mute. The triangle was ruptured.

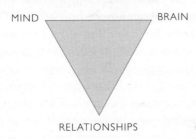

THE TRIANGLE OF WELL-BEING

MIND — BRAIN

RELATIONSHIPS

SEEING CLEARLY, LETTING GO, LETTING IN

I met with Leanne, Amy, Tommy, and Ben many times to give them an opportunity to talk openly to me and one another about how their lives had changed since Barbara's accident. Then one day I brought in Barbara's brain scans and pointed to the areas that had been damaged. I made simplified sketches on a whiteboard so that they could visualize the varied connections of the prefrontal cortex, and I let them know that the injury to this key region could explain almost all of the ways in which Barbara had changed. This seemed particularly important because children so often feel guilty when things go bad in a family. Here was concrete evidence that their mother's irritability and lack of warmth towards them were not caused by anything they had done and could not be remedied by their behaving "better." I hoped that instead of becoming paralyzed with self-recrimination or confusion, they could make sense of the change in their lives and experience directly the pain of their loss.

The children listened as attentively as Ben did, and even Tommy seemed to grasp that his mother had a "broken brain." Leanne had already become much more talkative during our meetings, and now she asked many questions about why her mum's love needed a brain to become "alive." "I thought love came from the heart," she said. She was right: The networks of nerve cells around the heart and throughout the body communicate directly with the social parts of our brain and they send that heartfelt sense right up to our middle prefrontal areas. I told Leanne that unless her mother's brain was working properly, she couldn't pick up the signals that I was sure were still there in her heart. That image seemed to soothe Leanne, and she came back to it later again and again. It gave her a new patience and tolerance for her mother's distant and irritable way of being, and I was touched to see her quiet acts of kindness towards Barbara. Leanne began talking again in school, reconnected with her friends, and found comfort in her teacher, who paid extra attention to her after hearing what had happened at home.

I met with Ben separately and encouraged him to express his own feelings more openly. This did not come naturally to him, and he had been working hard to keep family life as "normal" as possible. But of course their life wasn't normal, and the children needed to see that they were not alone in their grief, that it was okay for them to express their fears and concerns and uncertainties. Ben and I also discussed Tommy's particular needs. He had in essence lost his mother at two, before his own prefrontal region had begun to blossom. Not yet having developed the circuitry to express his feelings fully, Tommy especially would need ongoing help to make sense of his life's story. For now, at three, his sadness, anxiety, and confusion were almost beyond words.

Amy continued to struggle with her anger towards her mother. She was furious that Barbara had not worn a seat belt that day and frustrated that the mother she'd once looked up to was now gone. In addition, at the very time when she was starting to move away from her family and find her own identity with friends, she was expected to take care of Leanne and Tommy. I heard her frustration, and I helped Ben acknowledge her need to have time for herself even while still being expected to pitch in at home. Gradually she became able to treat her mother with more kindness, although Barbara could not reciprocate and treat Amy that way. This was their new reality.

As time went on, Barbara's motor coordination improved somewhat, but the damage to the front of her brain had been too severe, and she showed no signs of regaining her connected way of being. Nonetheless, Leanne and her family continued to strengthen their connections with one another. Mindsight permitted them to make sense of their experience and to allow the grieving process to unfold in a healthy way. Mindsight is what Barbara had lost—and mindsight was what the family needed to mourn the old Barbara and let the new Barbara into their lives.

I learned then that knowing about the different functions of the brain somehow enables people to gain enough distance from a damaged or hurtful relationship that they can develop more compassion and understanding, both for the other person in the relationship and for themselves. As you will see throughout this book, this lesson has guided my work as a therapist ever since.

Minding the Brain
The Brain in the Palm of Your Hand

MINDSIGHT DEPENDS UPON LINKING together wide arrays of neural input—from throughout the entire body, from multiple regions of the brain, and even from the signals we receive from other people. To understand how this linkage takes place, it helps to be able to visualize the brain as a system of interconnected parts.

Since the time when I first sketched Barbara's prefrontal regions for Leanne and the rest of her family, I've experimented with a number of models that show the brain in three dimensions. Following is the one I've never forgotten to take with me to a lecture. You can use it as you read this book without even getting up from your chair. Of course it's simplified enough to make some neurologists eager for more details, but it has helped many of my patients develop the mindsight needed to make sense of their experiences.

HAND MODEL OF THE BRAIN

If you put your thumb in the middle of your palm and then curl your fingers over the top, you'll have a pretty handy model of the brain. (My kids can't stand that pun, either.) The face of the person is in front of the knuckles, the back of the head towards the back of your hand. Your wrist represents the spinal cord, rising from your backbone, upon which the brain sits. If you lift up your fingers and raise your thumb, you'll see the inner brainstem represented in your palm. Place your thumb back down and you'll see the approximate location of the limbic area (ideally we'd have two thumbs, left and right, to make this a symmetric model). Now curl your fingers back over the top, and your cortex is in place.

These three regions—the brainstem, the limbic area, and the cortex—comprise what has been called the "triune" brain, which

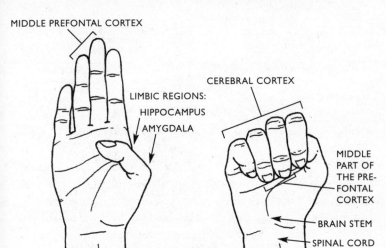

MIDDLE PREFONTAL CORTEX

LIMBIC REGIONS:

HIPPOCAMPUS

AMYGDALA

CEREBRAL CORTEX

MIDDLE PART OF THE PRE-FONTAL CORTEX

BRAIN STEM

SPINAL CORD

Place your thumb in the middle of your palm as in this figure.

Now fold your fingers over your thumb as the cortex is folded over the limbic areas of the brain.

developed in layers over the course of evolution. At a very minimum, integrating the brain involves linking the activity of these three regions. Since they are distributed bottom to top—from the inward and lower brainstem region, to the limbic area, to the outer and higher cortex—we could call this "vertical integration." The brain is also divided into two halves, left and right, so neural integration must also involve linking the functions of the two sides of the brain. This could be called "horizontal" or "bilateral integration." (I'll discuss bilaterality in chapter 6.) Knowing about the functions of the major regions of the brain can help you to focus your attention in ways that will create the desired linkage among them. So allow me to give you a brief overview of the layers of the triune brain.

THE BRAINSTEM

Hundreds of millions of years ago, the brainstem formed what some call the "reptilian brain." The brainstem receives input from the body and sends input back down again to regulate basic processes

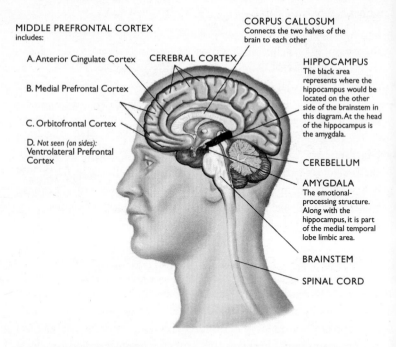

MIDDLE PREFRONTAL CORTEX
includes:

A. Anterior Cingulate Cortex

B. Medial Prefrontal Cortex

C. Orbitofrontal Cortex

D. *Not seen (on sides):* Ventrolateral Prefrontal Cortex

CEREBRAL CORTEX

CORPUS CALLOSUM
Connects the two halves of the brain to each other

HIPPOCAMPUS
The black area represents where the hippocampus would be located on the other side of the brainstem in this diagram. At the head of the hippocampus is the amygdala.

CEREBELLUM

AMYGDALA
The emotional-processing structure. Along with the hippocampus, it is part of the medial temporal lobe limbic area.

BRAINSTEM

SPINAL CORD

A diagram of the human brain looking from the middle to the right side. Some of the major areas of the brain are indicated, including the brainstem, the limbic areas (with the amygdala and hippocampus), and the cerebral cortex (with the middle prefrontal regions). Not seen is the ventrolateral prefrontal cortex.

such as the functioning of our heart and lungs. Beyond controlling the energy levels of the body through regulating heart rate and respiration, the brainstem also shapes the energy levels of the brain areas above it, the limbic and cortical regions. The brainstem directly controls our states of arousal, determining, for example, if we are hungry or satiated, driven by sexual desire or relaxed with sexual satisfaction, awake or asleep.

Clusters of neurons in the brainstem also come into play when certain conditions seem to require a rapid mobilization of energy distribution throughout the body and brain. This so-called fight-flight-freeze array of responses is responsible for our survival at times of danger. Working in concert with the evaluative processes of both the limbic and the higher cortical regions, the brainstem is the arbiter of whether we respond to threats either by mobilizing our energy for combat or for flight, or by freezing in helplessness, collapsing in the face of an overwhelming situation. But whichever of these responses is chosen, when we are in survival mode our reactivity makes it quite challenging, if not outright impossible, to be open and receptive to others. So part of the process of developing mindsight involves reducing reactivity when it's not actually necessary, as you will see later.

The brainstem is also a fundamental part of what are called "motivational systems" that help us satisfy our basic needs for food, shelter, reproduction, and safety. When you feel a deep "drive" to behave in a certain way, chances are that your brainstem is working closely with the next-higher region, the limbic area, to push you to act.

THE LIMBIC REGIONS

The limbic area lies deep within the brain, approximately where your thumb is on the hand model. It evolved when small mammals first appeared around two hundred million years ago. This "old mammalian brain" works closely with the brainstem and the body proper to create not only our basic drives but also our emotions. These feeling states are filled with a sense of meaning because the limbic regions evaluate our current situation. "Is this good or is this bad?" is the most basic question the limbic area addresses. We move towards the good and withdraw from the bad. In this way the limbic regions help create the "e-motions" that "evoke motion," that motivate us to act in response to the meaning we assign to whatever is happening to us in that moment.

The limbic area is also crucial for how we form relationships and become emotionally attached to one another. If you've ever raised fish, or frogs, or lizards, you know that these nonmammalian creatures lack attachment to you—and to one another. Rats, cats, and dogs, on the other hand, are equipped with a mammalian limbic region. Attachment is just what they—and we—do. We are hardwired to connect with one another thanks to our mammalian heritage.

The limbic area plays an important regulatory role through the hypothalamus, a master endocrine control center. Via the pituitary gland, the hypothalamus sends and receives hormones throughout the body—especially influencing our sexual organs and the thyroid and adrenal glands. For example, when we are stressed we secrete a hormone that stimulates the adrenals to release cortisol, which mobilizes energy by putting our entire metabolism on high alert to meet the challenge. This response is highly adaptive in the face of short-term stress, but it can turn into a problem in the long term. If we face an overwhelming situation in which we cannot adequately cope, cortisol levels may become chronically elevated. Traumatic experiences, in particular, can sensitize limbic reactivity, so that even minor stresses can cause cortisol to spike, making daily life more challenging for the traumatized person. These high cortisol levels can also be toxic to the growing brain and interfere with proper growth and function of neural tissue. Finding a way to soothe excessively reactive limbic firing is crucial to rebalancing emotions and diminishing the harmful effects of chronic stress. As we'll see, mindsight can help us recruit the higher areas of the brain to create a "cortical override" of these limbic reactivities.

The limbic area also helps us create several different forms of memory—of facts, of specific experiences, of the emotions that gave color and texture to those experiences. Located to either side of the central hypothalamus and pituitary, two specific clusters of neurons have been intensively studied in this regard: the amygdala and the hippocampus. The almond-shaped amygdala has been found to be especially important in the fear response. (Although some writers attribute all emotions to the amygdala, more recent research suggests that our general feelings actually originate from more broadly distributed areas of the limbic zone, the brainstem, and the body proper, and are woven into our cortical functioning as well.)

The amygdala can prompt an instantaneous survival response. Once, when my son and I were hiking in the High Sierra, a sudden jolt of fear brought me to a halt and I yelled out to him, "Stop!" Only after I yelled did I realize why—my constantly on-guard amygdala had seized upon a visual perception, beneath my conscious awareness, of a coiled object in our path. Luckily my son did stop (he wasn't yet a resistant teenager) and was then able to step around the poised-for-action young rattlesnake who was sharing the trail with us. Here we see that emotional states can be created without consciousness, and we may act on them without awareness. This may save our lives—or it can cause us to do things we later regret deeply. In order for us to become aware of the feelings inside us—to consciously attend to and understand them—we need to link these subcortically created emotional states to our cortex.

Finally we come to the hippocampus, a sea horse–shaped cluster of neurons that functions as a master "puzzle-piece-assembler," linking together widely separated areas of the brain—from our perceptual regions to our repository for facts to our language centers. This integration of neural firing patterns converts our moment-to-moment experiences into memories. I can relate the snake story to you because my hippocampus linked together the various aspects of that experience—sensations in my body, emotions, thoughts, facts, reflections—into a lived-in-time set of recollections.

The hippocampus develops gradually during our early years and continues to grow new connections and even new neurons throughout our lives. As we mature, the hippocampus weaves the basic forms of emotional and perceptual memory into factual and autobiographical recollections, laying the foundation for my ability to tell you about that long-ago snake encounter in the Sierras. However, this uniquely human storytelling ability also depends upon the development of the highest part of the brain, the cortex.

THE CORTEX

The outer layer, or "bark," of the brain is the cortex. It is sometimes called the "new mammalian" brain or neocortex because it expanded greatly with the appearance of primates—and most especially with the emergence of human beings. The cortex creates more intricate firing patterns that represent the three-dimensional world beyond the bodily functions and survival reactions mediated by the lower, subcortical regions. In humans, the more elaborate frontal portion of the cortex allows us to have ideas and concepts and to develop the mindsight maps that give us insight into the inner world. The frontal cortex actually makes neural firing patterns that represent its own representations. In other words, it allows us to think about thinking. The good news is that this gives us humans new capacities to think—to imagine, to recombine facts and experiences, to create. The burden is that at times these new capacities allow us to think too much. As far as we know, no other species represents its own neural representations—probably one reason why we sometimes call ourselves "neurotic."

The cortex is folded into convoluted hills and valleys, which brain scientists have divided into regions they call lobes. On your hand model, the back or posterior cortex extends from your second knuckle (counting from the fingertips) to the back of your hand, and includes the occipital, parietal, and temporal lobes. The posterior cortex is the master mapmaker of our physical experience, generating our perceptions of the outer world—through the five senses—and also keeping track of the location and movement of our physical body through touch and motion perception. If you've learned to use a tool—whether it was a hammer, a cricket bat, or even a car—you may remember the magical moment when your initial awkwardness dropped away. The amazingly adaptive perceptual functions of the back of the cortex have embedded that object into your body-maps so that it is neurally experienced like an extension of your body. This is how we can drive rapidly on a motorway or park a car in a tight space, use a scalpel with precision, or hit a six.

Looking again at your hand model, the front of the cortex, or frontal lobe, extends from your fingertips to the second knuckle. This region evolved during our primate history and is most developed in our human species. As we move from the back towards the front, we first encounter a "motor strip" that controls our voluntary muscles. Distinct groups of neurons control our legs, arms, hands, fingers, and facial muscles. These neural groups extend to the spinal cord, where they cross over, so that we make our right-side muscles work by activating our left motor area. (The same crossover is true for our sense of touch, which is represented farther back in the brain, in a zone of the parietal lobe called the "somatosensory strip.") Coming back to the frontal area and moving a bit more forward, we find a region called the "premotor" strip, which allows us to plan our motor actions. You can see that this part of the frontal lobe is still deeply connected to the physical world, enabling us to interact with our external environment.

THE PREFRONTAL CORTEX

As we move higher and more forward in the brain, we finally come to the area from your first knuckles to your fingertips on the hand model. Here, just behind the forehead, is the prefrontal cortex, which has evolved to this extent only in human beings. We have now moved beyond the neural concerns for the physical world and the movement of the body and into another realm of neurally constructed reality. Beyond the bodily and survival concerns of the brainstem, beyond the evaluative and emotional limbic functions, beyond even the perceptual processes of the posterior cortex and the motor functions of the posterior portion of the frontal lobe, we come upon the more abstract and symbolic forms of information flow that seem to set us apart as a species. In this prefrontal realm, we create representations of concepts such as time, a sense of self, and moral judgments. It is here also that we create our mindsight maps.

Look again at your hand model. The outer two fingertips represent the side prefrontal cortex, which participates in generating the conscious focus of attention. When you put something in the "front of your mind" you are linking activity in this region to activity from other areas of the brain, such as the ongoing visual perceptions from the occipital lobe. (Even when we generate an image from memory, we activate a similar portion of that occipital lobe.) When my amygdala perceived the rattlesnake without my conscious awareness, that perceptual "shortcut" likely took place without my side prefrontal involvement. Only later, after I'd yelled for my son to stop and felt my heart pounding, did my side prefrontal region get involved and permit me to work out, consciously, that I'd been afraid of a snake.

Now focus on the middle two fingernail areas. We have arrived at the middle prefrontal area that was so severely damaged in Barbara's accident. As I described earlier in this chapter, this area has important regulatory functions that range from shaping bodily processes—through overseeing brainstem activity—to enabling us to pause before we act, have insight and empathy, and enact moral judgments.

What makes this middle prefrontal region so crucial to carrying out these essential functions of a healthy life? If you lift your fingers

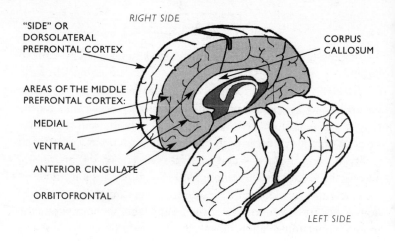

"SIDE" OR DORSOLATERAL PREFRONTAL CORTEX

AREAS OF THE MIDDLE PREFRONTAL CORTEX:

MEDIAL

VENTRAL

ANTERIOR CINGULATE

ORBITOFRONTAL

RIGHT SIDE

CORPUS CALLOSUM

LEFT SIDE

The two halves of the brain. This figure reveals the locations of the areas of the middle prefrontal cortex, which includes the medial and ventral regions of the prefrontal cortex, the orbitofrontal cortex, and the anterior cingulate cortex on both sides of the brain. The corpus callosum connects the two halves.

up and put them back down, you'll get a sense of the anatomical uniqueness of this region: It connects everything. Notice how your two middle fingertips rest on top of the limbic-thumb and touch the brainstem-palm, and are also linked directly to the cortex-fingers. So the middle prefrontal area is literally one synapse away from neurons in the cortex, the limbic area, and the brainstem. And, as I'll discuss later, it even has functional pathways that connect us to the social world of other brains.

The middle prefrontal region creates links among the following widely separated and differentiated neural regions: the cortex, limbic areas, and brainstem within the skull, and the internally distributed nervous system of the body proper. It also links signals from all those areas to the signals we send and receive in our social world. As the prefrontal cortex helps coordinate and balance the firing patterns from these many regions, it is profoundly integrative.

In the following chapter we'll explore what happens when this integrative area goes off-line. Lift up your fingers and you'll have an image of how we "lose it" and head down the "low road" in our interactions with others.

2

Crepes of Wrath
Mindsight Lost and Found

WHEN THE MIND WORKS WELL, when our brain functions as an integrated whole, our relationships thrive. But sometimes we "lose our minds" and act in ways we do not choose. The story I'll share with you in this chapter was a lesson in impaired mindsight—and a reminder that no matter how hard you try, you are only human and your mind will remain full of vulnerabilities and rough spots.

It was a warm spring day and I walked with my nine-year-old daughter on an open pedestrian promenade to find her brother. She and I had just seen a delightfully funny film and she was skipping along the sidewalk as we looked around the busy street. My lanky thirteen-year-old son had gone with some of his classmates to another cinema, but now he spotted us, waved, and left his friends to join us. On the way back to the car, we passed a crepe shop and he asked if we could stop. We had time before we had to get home and so we stepped into the small shop for a snack.

My son ordered a small crepe for himself, but my daughter said she wasn't hungry. The crepe arrived, aromas wafting from the open kitchen behind the counter where my son had placed his order. We sat down and he took his first forkful. Then my daughter asked if she could try some. My son looked at the small crepe and said that he was hungry and she could order her own. Reasonable suggestion, I thought, so I offered to get another crepe for her—but she said she only wanted a small bite to see how it tasted. That also seemed reasonable, so I suggested that my son share a piece with his sister.

If you have more than one child at home, or if you've grown up with a brother or sister, you may be very familiar with the game of sibling chess, strategic interactions filled with ever-present sets of movements aimed to assert power and achieve parental recognition and approval. But even if this were not such a sibling-assertion game, the small cost of buying a second crepe would have been quite a simple one to pay to avoid what you may guess was about to happen. Instead of making the purchase, I made a parental blunder and took sides. I firmly asserted that my son share his crepe with his sister. If this was not sibling chess before, it certainly became that after I stepped into their interaction.

"Why don't you just give her a small piece so she can see how it tastes," I urged.

He looked at me, at his crepe, and then with a sigh he gave in. Even as a young teenager he was still listening to me. Then, using his knife like a scalpel, he extracted the smallest piece of crepe you can imagine, one you'd almost need tweezers to pick up. Under other circumstances, I might have laughed and seen this as a creative sibling-chess move.

My daughter took the specimen, placed it on her napkin, and said that it was too small. Another great move.

In just a beat of time he responded, without looking up from his plate, that she couldn't be choosy. The chess match was on, full force, and I couldn't see it.

Although I knew that teens and their younger siblings don't get along all the time, that they frequently engage in creative and varied chess matches, in subtle or not so subtle ways, their interaction was getting to me.

Now I was beginning to boil. "Can you give her a real piece, one that you can actually see?" He excised another, larger piece and I felt relieved.

Then my daughter complained that this was the burned part—and sure enough, he had cut off that burned corner of the crepe that crumbles tasteless in your mouth. Chess moves well executed.

An outsider looking in at us at the table might have seen nothing out of the ordinary: a dad and his two animated kids out for some food. But inside, I was about to explode. When the bantering continued, something inside of me shifted. My head began to spin, but I told myself that I'd remain calm and appeal to reason. I could feel my face tense up, my fists get taut, and my heart begin to beat faster, but I tried to ignore these signals. That was it for me. Feeling overwhelmed by the ridiculousness of the whole encounter, I got up, took my daughter's hand, and went outside to wait on the sidewalk in front of the shop until my son finished his crepe. A few minutes later he emerged and asked why we had left. As I stormed off towards the car, my daughter in tow and my son hurrying to keep up, I told them that they should learn to share their food with each other. He pointed out in a matter-of-fact tone that he did give her a piece, but by then I was boiling over with frustration, and at that point there was no turning off the heat under the kettle. We got to the car and, fired up, I ignited the engine and away we went towards home. They had been normal siblings out for a film and a snack. I became a father out of my mind.

I couldn't let it go. Sitting next to me in the passenger seat, my son just countered everything I came up with by some rational, measured response as any teenager would do. In fact, he seemed quite adept at staying calm as he dealt with his now-irrational father. In that state, I just became more irate and gave him inappropriate consequences for things he didn't even do.

WHEN WE LOSE OUR MIND

I'm not proud to tell you any of this. But I do feel that since such explosive episodes are quite common, it is essential that we acknowledge their existence and help one another understand how mindsight can diminish their negative impact on our relationships and on our world. In our shame, we often try to ignore that a meltdown has occurred. But if we own the truth of what has happened, not only can we begin to repair the damage—which can be quite toxic to ourselves as well as to others—we can also actually decrease the intensity of such events and the frequency with which they occur.

Let's look at my meltdown again in terms of how my mind was riding the waves of my (mis)firing brain. A likely explanation is that I experienced a temporary brain dysfunction similar to what I described in chapter 1 when discussing the sudden irrational emotional eruptions that Barbara experienced following her accident.

In that kind of dysfunction, under certain crepelike conditions, the "limbic lava" from the fiery emotional centers below the cortex, just beneath the middle prefrontal area, can explode in out-of-control activity. All sorts of factors can contribute to such a meltdown, including lack of sleep and hunger—which were both true in my case on that day—and the particular meaning of an event, as we'll soon discover. The middle prefrontal cortex—the region that calms the emotionally reactive lower limbic and brainstem layers—stops being able to regulate all the energy being stirred up, and the coordination and balance of the brain is disrupted. That's my understanding of what happens when we go down the "low road," moving directly from limbic impulse to speech and action, and detouring away from the prefrontal "high road," where we are flexible and receptive rather than inflexible and reactive. We "lose it."

Without the prefrontal cortex's capacity to create you-maps, I was unable to see my son's and daughter's behavior as sibling-chess moves in an evolving sequence of bids for recognition and power. Nothing out of the ordinary, if you see the mind behind the behavior. Without me-maps, I could not see the meaning of the interaction for my own symbol-making mind, echoes of a past, as we'll soon discover. And without the we-maps, I could not see the inappropriate parental response of intervening with a teen and preteen negotiating their own sibling relationship. This intervention actually intensified the banter into an argument, propelled forward by my own emotional reactions. I became an inadvertent participant in their perhaps unintentional game of sibling chess.

THE MECHANISM OF MINDLESSNESS

Let me briefly map my meltdown against the nine prefrontal functions that I introduced in chapter 1. To summarize, they are 1) bodily regulation, 2) attuned communication, 3) emotional balance, 4) response flexibility, 5) fear modulation, 6) empathy, 7) insight, 8) moral awareness, and 9) intuition. These nine would top many researchers' and therapists' lists of the elements of emotional well-being. This is also a list of what I lost when I lost my mind.

Bodily regulation. The middle prefrontal region coordinates the activity of a part of the nervous system that controls bodily functions such as heart rate, respiration, and digestion. This "autonomic" nervous system has two branches: the *sympathetic,* which is often compared with a car's accelerator, and the *parasympathetic*—the brakes. Balancing the two allows us to drive the car of the body smoothly, so that we lift up off the brakes when we press the accelerator, and vice versa. Without such coordination, we can burn out, revving up while trying to slow down.

During my meltdown, my heart was beating out of control and my intestines were churning—just as if I'd been facing a physical threat.

Attuned communication. When we attune to others we allow our own internal state to shift, to come to resonate with the inner world of another. This resonance is at the heart of the important sense of "feeling felt" that emerges in close relationships. Children need attunement to feel secure and to develop well, and throughout our lives we need attunement to feel close and connected.

When I went down the low road, I could no longer attune to my children; I was unable to align my own state with theirs.

Emotional balance. When we are in emotional balance, we feel alive and at ease. Our feelings are aroused enough for life to have meaning and vitality, but not so aroused that we feel overwhelmed or out of control. Lacking balance, we move towards either excessive arousal, a state of chaos, or too little arousal, a state of rigidity or depression. Either extreme drains us of vitality. In the face of life's challenges, even the healthiest person may be temporarily "thrown off " and feel out of balance, but the middle prefrontal region functions to bring us back to equilibrium. This is the brain basis of equanimity, the ability to stay clear and focused in the face of storms from both inside and outside of us.

I lost my equanimity somewhere between the third and fourth round of banterings about sharing the crepe.

Response flexibility harnesses the power of the middle prefrontal region to put a temporal space between input and action. This ability to pause before responding is an important part of emotional and social intelligence. It enables us to become fully aware of what is happening—and to restrain our impulses long enough to consider various options for response. We work hard to model and to teach this to our children, and we can continue to strengthen this ability throughout the lifespan.

Early in the crepe episode, I felt fine. But then I became aware that something had changed inside of me, and relatively rapidly a state of chaotic agitation arose and made me inflexible. Trapped by my rising anger, I became unable to pause before speaking or acting.

Fear modulation. After experiencing a frightening event, we may come to feel fear in the face of a similar situation. But the middle prefrontal region has direct connections that pass down into the limbic area and make it possible to inhibit and modulate the firing of the fear-creating amygdala. Studies have demonstrated that we can consciously harness this connection to overcome fear—we can use the "override" of our cortex to calm our lower limbic agitation. (After we had discussed the role of the brain in her treatment, one of my young patients announced, "I'm trying to get my prefrontal cortex to squirt GABA-goo over my amygdala." GABA, or gamma-aminobutyric acid, is a neurotransmitter that plays an important role in the prefrontal inhibition of subcortical firing, and she had imagined it as a kind of gel to soothe the limbic eruptions.)

As I realized later, my irritation and ensuing anger were indeed driven by an old fear—one I had worked hard to understand and master (more about that when I take up the story again). But all those gains were now temporarily suspended, and GABA-goo was not at my command, dried up by the heat of my rage.

Empathy is the capacity to create mindsight images of other people's minds. These you-maps enable us to sense the internal mental stance of another person, not just to attune to their state of mind. Attunement is important, but the middle prefrontal cortex also moves us from this resonance and feeling-with to the more complex perceptual capacity to "see" from another's point of view: We sense the other's intentions and imagine what an event means in his or her mind.

Insight allows us to make me-maps enabling us to perceive our own mind. This creates what one researcher calls "mental time travel," in which we connect the past to the present and the anticipated future. The middle prefrontal region plays a crucial role in this mental time travel, permitting us to experience ourselves as a center of subjective gravity, the author of our own unfolding life story.

Both empathy and insight were casualties of my meltdown. I lost insight into my own mind, and I could not put myself in the place of either my son or my daughter, or even pause to wonder what they might be thinking and feeling. Without these maps, as we've seen, I could not have perspective on the minds beneath the behavior.

Moral awareness as I'm using it here denotes the ways in which we both think about and enact behaviors for the social good, and we have evidence that it requires an intact middle prefrontal region. Functional magnetic resonance imaging scanners have shown that that region becomes highly active when we imagine actions for the larger social good. Other research has shown that when the middle prefrontal region is damaged, we may become amoral. Moral reasoning seems to require the integrative capacity of this region of the brain both to sense the emotional meaning of present challenges and to override immediate impulses in order to create moral action in response to those challenges. This may be how the we-maps created by the middle prefrontal cortex enable us to move beyond our immediate, individually focused survival needs, and even beyond the present version of our relationship maps, to a vision of a larger, interconnected whole.

From a moral perspective, my meltdown included unfairly focusing on my son and imposing unreasonable, even ridiculous consequences that had nothing to do with the "larger good" for all concerned. I was driven by my own personal feelings and reactivity, not a sense of what was right or fair.

Finally, *intuition* can be seen as how the middle prefrontal cortex gives us access to the wisdom of the body. This region receives information from throughout the interior of the body, including the viscera—such as our heart and our intestines—and uses this input to give us a "heartfelt sense" of what to do or a "gut feeling" about the right choice. This integrative function illuminates how reasoning, once thought to be a "purely logical" mode of thinking, is in fact dependent on the nonrational processing of our bodies. Such intuition helps us make wise decisions, not just logical ones.

But with my limbic ball of fire burning, I had no access to intuition—to the wisdom of my body and to a deeper sense of knowing what was true, what was really happening. Paradoxically, however, I might have said that I felt justified in what I was doing, that "in my gut" it felt right. Those statements would have been rationalizations driven by my increasing irritation and consistent with my rising anger and agitated voice.

Though all of this is embarrassing to relate, I offer it as evidence that we are all potentially prone to such low-road disintegrations. The key is recognizing when they happen, putting an end to them as quickly as possible to minimize the hurt they cause, and then making a repair. We need to regain what was truly lost—mindsight—and then use mindsight to reconnect with ourselves and with those for whom we care so deeply.

MAKING SENSE OF A MELTDOWN

On the day of the crepes, I was still irritated with my son when we entered the house. I walked into another room, away from my son, took a deep breath, stretched, and tried to calm down. I knew repair was crucial, but my vital signs were through the roof and I had to bring them into balance before I could do anything else.

I knew that being outside and doing something physical would help, so I went with my daughter for a roller-skating break around the neighborhood, one of our favorite activities together ever since she was six years old. We skated for a while in silence, hand in hand. I could sense the rhythm of our moving together and feel the air against my body as we glided down the street. I was literally starting to come to my senses.

After a while my daughter asked me why I had yelled at her brother, just about a crepe.

Good question. I told her I thought sharing was important (lame excuse, I know, but that's what I thought at the time).

At that moment, I felt a flow of associations rise up in my mind like the pages of a photo album from my childhood, the pictures flashing by in front of my eyes. I came to realize that what happened is that I saw my daughter as a symbol of myself when I was a child, and my son as a symbol of my older brother when he was a teen. I had images of my brother playing with me when we were young, and even protecting me from other kids when we both were in elementary school. But when he became a teenager, we no longer got along so well and rarely shared time together. Even though he and I are close as adults and we laugh about those days now, back then it was a painful time for me. I told my daughter as we skated along that I had decided that if I ever had kids I'd try to make sure they got along.

Then—most insightfully—my daughter offered that this was my issue, not hers or her brother's. She even said that I should work it out on my time, not through them.

She was right, of course. As we skated together, my mind now calm and my prefrontal region back online, I could begin to reflect on what had been going on. I now could look within myself at the emotions that had erupted, and I could see the issues that had contributed to my meltdown.

What had happened during my roller-skating break that allowed me to regain mindsight?

THE TRIPOD OF REFLECTION: OPENNESS, OBSERVATION, OBJECTIVITY

To regain control of the mind after we have lost it, we need the power of reflection that is at the heart of mindsight. Mindsight emerges as our communication—with others and with ourselves—helps us reflect on who we really are and what is going on inside us. Here I'll explore three very specific components of reflection that are at the heart of our mindsight abilities: openness, observation, and objectivity.

I like to think of these fundamental components as the three legs of a tripod that stabilizes our mindsight lens. Without the tripod, our mind may be visible to us only as a blurry, busy hive of activity whose fine details are lost in jumping images and fleeting feelings. But when the lens of our mindsight camera is stabilized, the details come into focus. We see with more depth and precision. From this stabilization we gain all the gifts of acuity: keenness, insight, perception, and, ultimately, wisdom.

Openness implies that we are receptive to whatever comes to our awareness and don't cling to preconceived ideas about how things "should" be. We let go of expectations and receive things as they are, rather than trying to make them how we want them to be. Openness enables us to sense things clearly. It gives us the power to recognize restrictive judgments and release our minds from their grip.

Observation is the ability to perceive the self even as we are experiencing an event. It places us in a larger frame of reference and broadens our perspective moment to moment. Put another way, self-observation allows us to see the fuller context in which we are living. Observation offers a powerful way to disengage from automatic behaviors and habitual responses; we can sense our role in these patterns and begin to find ways to alter them.

Objectivity permits us to have a thought or feeling and not become swept away by it. It recruits the ability of the mind to be aware that its present activities—our thoughts, feelings, memories, beliefs, and intentions—are temporary and, moreover, that they are not the totality of who we are. They are not our identity. Objectivity allows us to develop what is sometimes called discernment. With discernment we can see that a thought or feeling is just mental activity, not absolute reality. Later in the book we'll explore this ability in more detail, but here let me just mention that one part of discernment is the ability to be aware of how we are being aware—as opposed to becoming lost in the target of our attention. This "meta-awareness," or awareness of awareness, is a powerful skill that can liberate us from the prison of automatic reactions.

So the essence of reflection, which is central to mindsight, is that we remain open, observant, and objective about what's going on both inside us and inside others. Without any one of these three tripod legs, mindsight becomes unsteady and our ability to clearly see the mind—of ourselves or of others—becomes compromised.

When I lost control over the crepes, I was being reactive rather than receptive. Had I remained open and reflective, I might have been able to make our interaction a time of learning for all of us. Instead I was carried away by the intensity of my emotions, my feelings overwhelmed my awareness, a subcortical storm disabled prefrontal integration, and my behavioral impulses went on autopilot.

Let's look at another, more neutral, example: listening to a piece of music. Naturally there are times when we "just listen" to music, get lost in it, and enter the "flow" of the melody. We are immersed, self-consciousness disappears, and the boundaries between ourselves and the focus of our attention—the music—melt away. Flow can be fabulous. But sometimes we absolutely need reflection and not flow. In many ways I was in the "flow" of my wrath at my son; I lost self-consciousness and became "one" with the fury. That is not a good thing, obviously. And so it's important to distinguish the reflection inherent to mindsight from the experience of flow. Reflection is crucial to pulling oneself out of the groove of a crepes-of-wrath experience—and also, later on, to making a repair. If we try to reconnect without reflection, if we simply revisit what happened, we can actually evoke the same reactive flow and fall back into the meltdown experience all over again.

But with reflection we can observe ourselves with openness and objectivity. We can sense the flood of out-of-control emotions as a mere part of the story of who we are. We gain the crucial capacity to deal with an intense emotion without becoming lost in it. This can make all the difference between explosion and expression.

No question, when we are in a meltdown, it's hard to recruit our reflective skills. But once we leave that disconnected, explosive state, reflection also helps us to look back and look inward at what has happened. If we recognize that this mental event was not the totality of who we are, we gain the reflective distance and the freedom to take responsibility for our actions and feelings. We can look at our autopilot behaviors and come to a deeper understanding that may permit us to behave differently in the future.

REFLECTION AND RECONNECTION

After the crepes episode, my daughter and I reconnected in our skating and talking together. I apologized to her for getting so upset. Now what I needed to do was to reconnect with my son.

When we are filled with out-of-control anger, we can't expect others to empathically say, "Oh, tell me more about how furious you are." Anger creates anger, and cooling off is essential before a repair process can be initiated. Even a short break can make all the difference. Then, if you value your relationship, it is crucial to take the initiative and make an effort to reconnect. This is especially true for parents. We parents are supposed to be the wiser, kinder, more mature persons, and even as we acknowledge that this is not always the case, it is at least a goal we can aim for. On the other hand, not beating up on ourselves is crucial to overcoming the shame and guilt we may feel after taking momentary leave of our sanity. Being kind to ourselves actually helps us to take the necessary steps towards repair and reconnection. It also helps to be prepared for being rebuffed initially, since this often happens when we first try to make a repair. Without such preparation, we may rapidly reenter the dis-integrated state, reinforcing and hardening the disconnection we are trying to undo.

Before we reconnect with others, we need to be sure we are connected with ourselves. To reconnect with myself, I needed to check in with the essential components of mental life—to reflect on my internal sensations, images, feelings, and thoughts. This reflection is like reviewing a checklist before you leave the house. Do you have your wallet, your keys, your calendar, your phone? Focusing on the internal life of the mind is often ignored in the hustle and bustle of everyday life.

Reviewing the crepes episode, I asked myself: What were the sensations in my body? What images did I have in my mind's eye? What were the feelings—the emotions—floating around in my head? What thoughts were occurring then—and are they still with me? In my meltdown state I was filled with the sensations of my tense body and pounding heart; images of my children arguing; feelings of anger and frustration; thoughts about how my son should be behaving. Now I could reflect on those experiences from a further distance, with an openness, observation, and objectivity I had lost at the time. I could also now see the deeper issues echoing in my memory that led to the meltdown.

Again, it would have been easy to beat up on myself: "What's wrong with you, Dan? You've written books on this subject, thought about this for years. . . . Why can't you keep your head together?" But reflection requires an attunement to the self that is supportive and kind, not a judgmental stance of interrogation and derogation. Reflection is a compassionate state of mind.

In many ways, I realized, I had lost those nine middle prefrontal functions. My low-road state likely resulted from the temporary shutting down of my middle prefrontal area. No longer in a state of integration, my brain lost its balance and coordination. The lower limbic, brainstem, and bodily areas held sway as my (usually) more reasonable, empathic, and flexible cortex went offline. Cooling down meant beginning to regain integration.

Once I had worked out what had been going on during the melt-down, what had triggered the fury and maintained it, I could also reflect on my mind to sense when I was feeling solidly enough on integrated ground that I could trust myself to be able to have a dialogue with my son. With the prefrontal cortex back in business, empathy had returned and I now began to focus on how important it was to repair our ruptured connections, and what I needed to do to make that possible.

SHIFTING TOWARDS REPAIR

When I finally cooled down after talking and skating and reflecting, I went to my son's room and asked if we could talk. I said that I thought I had gone off the deep end and that it would be helpful for us to discuss what had happened. He told me that he thought I was too protective of his sister. He was indeed, absolutely right. Although the embarrassment of having become irrational created an urge to speak up to defend myself and my reactions, I just kept quiet (*observation*). I could discern that this urge and the accompanying sensations were just activities of my mind, not the totality of who I was (*objectivity*). I did not have to speak just because the impulse was there. My son went on to tell me that my getting "upset" was unnecessary because he really hadn't done anything wrong. He was right. Again I felt a defensive urge to give him a lecture about sharing. But I reminded myself to remain reflective and focus on my son's experience, not mine. The essential stance here was to not judge who was right, and instead be accepting and recep-tive to him (*openness*). You can imagine that this all required mind-sight, for sure. I was thankful my prefrontal region was back at work.

With my daughter's question, I had already explored what I thought had been going on in me. I had come to realize that I had become swept up in my old leftover issues and I could no longer see clearly. Now I could just listen as he continued, without needing much guidance, to explain his point of view. I later offered to him that during the crepes encounter I had in fact taken his sister's side, unfairly, that I could see how this felt unjust to him, and that my explanation seemed irrational—because in fact it was. As an explanation—not an excuse—I let him know what had happened in my mind, seeing him as a symbol of my brother, so that we both could make sense of the whole encounter. Even though I probably looked awkward and clumsy in his teenage mind, I could tell that he knew my commitment to our relationship was deep and my effort to repair the damage was genuine. My mindsight had returned, our two minds connected again, and our relationship was back on track.

The key to the reflective dialogue I engaged in with my son was maintaining the three components of openness, observation, and objectivity. Each of these elements enables a powerful source of healing to emerge following a disruption in our relationships, and each of them is an essential part of the kindness we need to extend after such disruptions.

As I think back on the events of that day now, I realize once again how many layers of meaning our brains contain, and how quickly old, perhaps forgotten, memories can emerge to shape our behavior. These associations can make us act on automatic pilot. At the time of the crepes-of-wrath drama, that theme of feeling disconnected in my childhood was a "hot button" of leftover emotional issues in my life, ones that this incident made me realize I needed to reflect upon with more depth. With mindsight I was able to make use of the reflections that arose from that conflict to arrive at more clarifying insights into my own childhood experiences. This is how the most challenging moments in our lives can become opportunities to deepen our self-understanding and our connections with others.

As a wise professor of mine once said, "Uncovering memory and meaning are never over until life is over." He was so right. Even with intellectual understanding and reflective insights, we are still fallible, still human, still refining our mindsight skills. That day of crepes and cries and skates and insights had become part of our family's shared story. The repair process that we had engaged in following the turmoil had led not only to a repair but also to a deeper understanding for all of us. With mindsight our standard is honesty and humility, not some false ideal of perfection and invulnerability. We are all human, and seeing our minds clearly helps us embrace that humanity within one another and ourselves.

········

Minding the Brain
Neuroplasticity in a Nutshell

IT'S EASY TO GET OVERWHELMED thinking about the brain. With more than one hundred billion interconnected neurons stuffed into a small, skull-enclosed space, the brain is both dense and intricate. And as if that weren't complicated enough, each of your average neurons has ten thousand connections, or synapses, linking it to other neurons. In the skull portion of the nervous system alone, there are hundreds of trillions of connections linking the various neural groupings into a vast spiderweb-like network. Even if we wanted to, we couldn't live long enough to count each of those synaptic linkages.

Given this number of synaptic connections, the brain's possible on-off firing patterns—its potential for various states of activation—has been calculated to be ten to the millionth power—or ten times ten one million times. This number is thought to be larger than the number of atoms in the known universe. It also far exceeds our ability to experience in one lifetime even a small percentage of these firing possibilities. As a neuroscientist once said, "The brain is so complicated it staggers its own imagination." The brain's complexity gives us virtually infinite choices for how our mind will use those firing patterns to create itself. If we get stuck in one pattern or the other, we're limiting our potential.

Patterns of neural firing are what we are looking for when we watch a brain scanner "light up" as a certain task is being performed. What scans often measure is blood flow. Since neural activity increases oxygen use, an increased flow of blood to a given area of the brain implies that neurons are firing there. Research studies correlate this inferred neural firing with specific mental functions, such as focusing attention, recalling a past event, or feeling pain.

We can only imagine how a scan of my brain might have looked when I went down the low road in the crepes encounter: an abundance of limbic firing with increased blood flow to my irritated amygdala and a diminished flow to my prefrontal areas as they began to shut down. Sometimes, as happened that day, the out-of-control firing of our brain drives what we feel, how we perceive what is happening, and how we respond. Once my prefrontal region was off-line, the firing patterns from throughout my subcortical regions could dominate my internal experience and my interactions with my kids. But it is also true that when we're not traveling down the low road we can use the power of our mind to change the firing patterns of our brain and thereby alter our feelings, perceptions, and responses.

One of the key practical lessons of modern neuroscience is that the power to direct our attention has within it the power to shape our brain's firing patterns, as well as the power to shape the architecture of the brain itself.

As you become more familiar with the parts of the brain I discussed in the first Minding the Brain segment, you can more easily grasp how the mind uses the firing patterns in these various parts to create itself. It bears repeating that while the physical property of neurons firing is correlated with the subjective experience we call mental activity, no one knows exactly how this actually occurs. But keep this in the front of your mind: Mental activity stimulates brain firing as much as brain firing creates mental activity.

When you voluntarily choose to focus your attention, say, on remembering how London Bridge looked one foggy day last autumn, your mind has just activated the visual areas in the posterior part of your cortex. On the other hand, if you were undergoing brain surgery, the doctor might place an electrical probe to stimulate neural firing in that posterior area, and you'd also experience a mental image of some sort. The causal arrows between brain and mind point in both directions.

Keeping the brain in mind in this way is like knowing how to exercise properly. As we work out, we need to coordinate and balance the different muscle groups in order to keep ourselves fit. Similarly, we can focus our minds to build the specific "muscle groups" of the brain, reinforcing their connections, establishing new circuitry, and linking them together in new and helpful ways. There are no muscles in the brain, of course, but rather differentiated clusters of neurons that form various groupings called nuclei, parts, areas, zones, regions, circuits, or hemispheres. And just as we can intentionally activate our muscles by flexing them, we can "flex" our circuits by focusing our attention to stimulate the firing in those neuronal groups. Using mindsight to focus our attention in ways that integrate these neural circuits can be seen as a form of "brain hygiene."

WHAT FIRES TOGETHER, WIRES TOGETHER

You may have heard this before: As neurons fire together, they wire together. But let's unpack this statement piece by piece. When we have an experience, our neurons become activated. What this means is that the long length of the neuron—the axon—has a flow of ions in and out of its encasing membrane that functions like an electrical current. At the far end of the axon, the electrical flow leads to the release of a chemical neurotransmitter into the small synaptic space that joins the firing neuron to the next, postsynaptic neuron. This chemical release activates or deactivates the downstream neuron. Under the right conditions, neural firing can lead to the strengthening of synaptic connections. These conditions include repetition, emotional arousal, novelty, and the careful focus of attention. Strengthening synaptic linkages between neurons is how we learn from experience. And one reason that we are so open to learning from experience is that, from the earliest days in the womb and continuing into our childhood and adolescence, the basic architecture of the brain is very much a work in progress.

During gestation, the brain takes shape from the bottom up, with the brainstem maturing first. By the time we are born, the limbic areas are partially developed but the neurons of the cortex lack extensive connections to one another. This immaturity—the lack of connections within and among the different regions of the brain—is what gives us that openness to experience that is so critical to learning.

A massive proliferation of synapses occurs during the first years of life. These connections are shaped by genes and chance as well as experience, with some aspects of ourselves being less amenable to the influence of experience than others. Our temperament, for example, has a nonexperiential basis; it is determined in large part by genes and by chance. For instance, we may have a robust approach to novelty and love to explore new things, or we may tend to hang back in response to new situations, needing to "warm up" before we can overcome our initial shyness. Such neural propensities are set up before birth and then directly shape how we respond to the world—and how others respond to us.

But from our first days of life, our immature brain is also directly shaped by our interactions with the world, and especially by our relationships. Our experiences stimulate neural firing and sculpt our emerging synaptic connections. This is how experience changes the structure of the brain itself—and could even end up having an influence on our innate temperament.

As we grow, then, an intricate weaving together of the genetic, chance, and experiential input into the brain shapes what we call our "personality," with all its habits, likes, dislikes, and patterns of response. If you've always had positive experiences with dogs and have enjoyed having them in your life, you may feel pleasure and excitement when a neighbor's new dog comes bounding towards you. But if you've ever been severely bitten, your neural firing patterns may instead help create a sense of dread and panic, causing your entire body to shrink away from the pooch. If on top of having had a prior bad experience with a dog you also have a shy temperament, such an encounter may be even more fraught with fear. But whatever your experience and underlying temperament, transformation is possible. Learning to focus your attention in specific therapeutic ways can help you override that old coupling of fear with dogs. The intentional focus of attention is actually a form of self-directed experience: It stimulates new patterns of neural firing to create new synaptic linkages.

You may be wondering, "How can experience, even a mental activity such as directing attention, actually shape the structure of the brain?" As we've seen, experience means neural firing. When neurons fire together, the genes in their nuclei—their master control centers—become activated and "express" themselves. Gene expression means that certain proteins are produced. These proteins then enable the synaptic linkages to be constructed anew or to be strengthened. Experience also stimulates the production of myelin, the fatty sheath around axons, resulting in as much as a hundredfold increase in the speed of conduction down the neuron's length. And as we now know, experience can also stimulate neural stem cells to differentiate into wholly new neurons in the brain. This neurogenesis, along with synapse formation and myelin growth, can take place in response to experience throughout our lives. As discussed before, the capacity of the brain to change is called neuroplasticity. We are now discovering how the careful focus of attention amplifies neuroplasticity by stimulating the release of neurochemicals that enhance the structural growth of synaptic linkages among the activated neurons.

An additional piece of the puzzle is now emerging. Researchers have discovered that early experiences can change the long-term regulation of the genetic machinery within the nuclei of neurons through a process called *epigenesis*. If early experiences are positive, for example, chemical controls over how genes are expressed in specific areas of the brain can alter the regulation of our nervous system in such a way as to reinforce the quality of emotional resilience. If early experiences are negative, however, it has been shown that alterations in the control of genes influencing the stress response may diminish resilience in children and compromise their ability to adjust to stressful events in the future. The changes wrought through epigenesis will continue to be in the science news as part of our exploration of how experience shapes who we are.

In sum, experience creates the repeated neural firing that can lead to gene expression, protein production, and changes in both the genetic regulation of neurons and the structural connections in the brain. By harnessing the power of awareness to strategically stimulate the brain's firing, mindsight enables us to voluntarily change a firing pattern that was laid down involuntarily. As you will see throughout this book, when we focus our attention in specific ways, we create neural firing patterns that permit previously separated areas to become linked and integrated. The synaptic linkages are strengthened, the brain becomes more interconnected, and the mind becomes more adaptive.

THE BRAIN IN THE BODY

It's important to remember that the activity of what we're calling the "brain" is not just in our heads. For example, as I mentioned in chapter 1, the heart has an extensive network of nerves that process complex information and relay data upward to the brain in the skull. So, too, do the intestines, and all the other major organ systems of the body. The dispersion of nerve cells throughout the body begins during our earliest development in the womb, when the cells that form the outer layer of the embryo fold inward to become the origin of our spinal cord. Clusters of these wandering cells then start to gather at one end of the spinal cord, ultimately to become the skull-encased brain. But other neural tissue becomes intricately woven with our musculature, our skin, our heart, our lungs, and our intestines. Some of these neural extensions form part of the autonomic nervous system, which keeps the body working in balance whether we are awake or asleep; other circuitry forms the voluntary portion of the nervous system, which allows us to intentionally move our limbs and control our respiration. The simple connection of sensory nerves from the periphery to our spinal cord and then upward through the various layers of the skull-encased brain allows signals from the outer world to reach the cortex, where we can become aware of them. This input comes to us via the five senses that permit us to perceive the outer physical world.

The neural networks throughout the interior of the body, including those surrounding the hollow organs, such as the intestines and the heart, send complex sensory input to the skull-based brain. This data forms the foundation for visceral maps that help us have a "gut feeling" or a "heartfelt" sense. Such input from the body forms a vital source of intuition and powerfully influences our reasoning and the way we create meaning in our lives.

Other bodily input comes from the impact of molecules known as *hormones*. The body's hormones, together with chemicals from the foods and drugs we ingest, flow into our bloodstream and directly affect the signals sent along neural routes. And, as we now know, even our immune system interacts with our nervous system. Many of these effects influence the neurotransmitters that operate at the synapses. These chemical messengers come in hundreds of varieties, some of which—such as dopamine and serotonin—have become household names thanks in part to drug company advertising. These substances have specific and complex effects on different regions of our nervous system. For example, dopamine is involved in the reward systems of the brain; behaviors and substances can become addictive because they stimulate dopamine release. Serotonin helps smooth out anxiety, depression, and mood fluctuations. Another chemical messenger is oxytocin, which is released when we feel close and attached to someone.

Throughout this book, I use the general term *brain* to encompass all of this wonderful complexity of the body proper as it intimately intertwines with its chemical environment and with the portion of neural tissue in the head. This is the brain that both shapes and is shaped by our mind. This is also the brain that forms one point of the triangle of well-being that is so central to mindsight. By looking at the brain as an embodied system beyond its skull case, we can actually make sense of the intimate dance of the brain, the mind, and our relationships with one another. We can also recruit the power of neuroplasticity to repair damaged connections and create new, more satisfying patterns in our everyday lives.

3

Leaving the Ether Dome

Where Is the Mind?

WITHOUT MINDSIGHT, LIFE BECOMES DEADENED. When we find our-selves in a culture in which mindsight is absent, we can become stuck in the physical domain, blind to the internal reality at the heart of our lives. If the leaders of a culture are themselves devoid of mindsight, then the young, emerging minds of that culture will be living in a world in which the blind are leading the blind. Here I'd like to share with you the experience of a student immersed in such a mindsightless world, my introduction to the culture of modern medicine.

I first visited Harvard Medical School on a cold, gray winter day, and to a young man from Southern California, the bleakness only added to the authority of the huge stone buildings. Rigorous, demanding, challenging, Harvard was the mountain, and I wanted to climb it.

During my first two years, however, I was painfully and repeat-edly reprimanded for a peculiar interest of mine: spending time learning about my patients' life stories and inquiring about their feel-ings during patient interviews. I remember one report I made to a clinical supervisor. A sixteen-year-old African American boy seemed severely depressed by his diagnosis of sickle-cell anemia, and I discov-ered in talking with him that his older brother had died of the disease, after a long and excruciatingly painful decline, just four years earlier. Somehow no one had told him that his prognosis was much better—both because he had been diagnosed earlier than his brother, and because treatment had improved. He and I had been able to put words to the terrifying images of his brother's experience that were still in his mind, and together we created a more hopeful view of how things might work out for him.

My supervisor was a gastrointestinal specialist. "Daniel," she said, her head cocked to one side as if she thought I was lost or confused, "do you want to be a psychiatrist?"

"No," I said, "I'm just a second-year student and I have no idea what I want to do." Actually, I had been thinking about pediatrics as a specialty, since I loved being with children, but I wasn't going to mention that to her.

"Daniel," she said as she cocked her head to the other side, "is your *father* a psychiatrist?"

"No," I told her, "he's an engineer."

But that didn't seem to satisfy her, either. "You know these questions you are asking about the patients' feelings, about their lives? This is the work of social workers, not doctors. If you want to ask about those things, why don't you just become a social worker? If you want to be a real doctor, you need to stick to the physical."

My supervisor was telling me she wanted only the results of the physical exam, but in reality she was prescribing a worldview. She was not alone: The medical system at that time was focused almost exclusively on data and disease. Perhaps this was my teachers' way of dealing with the overwhelming feelings of facing illness and death every day, of feeling helpless at times, incompetent, or not in control. But to me their teaching seemed misguided and wrong. The patients' feelings and thoughts, their hopes and dreams and fears, their life stories, seemed just as real and important to me as their kidneys, or livers, or hearts. Yet there was no one—and no science—to show me a different way.

To survive those early years of medical indoctrination, I simply went along. I was young and eager to please my teachers, so I did the best I could to fit in with the system. I'm sure there must have been other students—and professors—who did not subscribe to the mindsightless worldview, but I couldn't find them. I once even tried to join the women's medical student organization, saying that I too needed humane role models. But I was told that men changed the dynamic in the room and I was politely but firmly asked not to intrude.

During my second year, my clinical rotation took place at Massachusetts General Hospital, and some of our classes were in the amphitheater where anesthesia had been introduced into modern medicine more than one hundred years before. I recall looking up at the dome that covered the hall, and staring blankly up into space and then down towards the far wall where a painting of that first surgical procedure hung in clear view of all the students. There the patient was, laid out cold on the table, numb to the feelings inside and oblivious to the black-coated men gathered around him. The hall was known as the Ether Dome, and I felt as if I too were being etherized—disconnected from my inner world, cut off from some living part of myself, and rapidly going unconscious. Even my body was going numb. I recall taking a shower and feeling nothing, and I had stopped going to the lively "Dance Free" nights every Wednesday night in a church across the river, something I had loved. I felt disengaged and lost. Dead.

Without quite understanding the reasons for my disillusionment, I called the dean of students and told her that I was dropping out of school. She listened kindly, and when she asked why I wanted to stop, I told her that I wasn't sure. I told myself that I had to leave to "find my sense of direction"; in reality, it was to find my own mind. The dean persuaded me to take a year's leave of absence instead, and she instructed me to write the "research request" necessary to justify my break. I wrote that I was going "to do research on who I was." Luckily there was a job opening for that position.

My "research" took me around the continent, from New England to British Columbia to Southern California. I tried out a number of careers, including professional dance and choreography, carpentry, and (almost) salmon fishing. I now imagine that the research I had done in college studying the molecular mechanisms salmon use to transition from fresh- to salt-water living was symbolic of some deeper interest in how we develop and change. On Vancouver Island, facing the wild Pacific on the western coast of British Columbia, I met a man who worked on the boats. Fishing, he told me, was all about "getting up at three A.M., bending over the side of a freezing boat for hours, your back killing you, throwing out fish-hooks, and pulling them in until your hands are crippled." Then he announced that he himself was quitting and going back to graduate school in psychology. That encounter sent me back to my hometown, where I reconnected with friends and family and helped my grandmother during the illness and death of my grandfather. Finally I got a job working with some documentary filmmakers who were taping the performing arts program at UCLA. They also asked me to assist with a research project about the left and right sides of the brain. That was it! I couldn't stop thinking about the mind, about our lives, about what makes us who we are. This was a path I could follow. I might become a psychiatrist after all. I felt ready to return to Harvard, and I was determined to remain—somehow—as clear and connected to myself and others as I'd felt during my year away.

NO TIME FOR TEARS

The capstone of my third year of medical school was the crucial clerkship in internal medicine. How well you did in that clerkship was reputed to determine your professional future. I was at a lecture when my supervising resident, a few years ahead of me in her training, came into the classroom, tears in her eyes, and whispered to me that Mr. Quinn, a patient I'd been caring for, had just died. I got up and went with her to his bedside. We stood there together for a long time. He had been a feisty merchant marine, his face roughened from years at sea. I used to sit with him after those long days at the hospital, soaking up his stories, listening to his feelings about his impending death. He knew that his seventy years on the planet were coming to an end, his adventures almost over. Now his life story was complete, and the resident and I shared our reflections as we stood by the body that had sailed his ship at sea.

That afternoon I met with the senior attending doctor for my mid-rotation student progress review. He was quite an imposing figure, tall, black-bearded, and handsome, an oncologist, who told me that I was doing a "fine job" in my clerkship—except for one thing. He noticed that I had left the teaching rounds that morning. I told him about Mr. Quinn's death and about how my resident and I had wanted to stay with him until the orderlies came to take his body away. Then the doctor said something I will never forget:

"Daniel, you have to realize that you are here to learn. Taking time away from a learning opportunity is a big problem. You have to get over these feelings—patients just die. There is no time for tears. Your job is to learn. To be an excellent doctor, you have to deal with just the facts."

No time for tears. Was this the art of medicine I was supposed to be learning?

The next day I went to Mr. Quinn's old room to admit a new patient. There I found one of my favorite science instructors sitting on the bed. He smiled at me and said, "Well, I guess these diseases can happen to any of us." He had developed acute leukemia, and I was supposed to begin preparing him for a bone marrow transplant. My face filled with intensity—first tears, which I held back; then fear, which I could not bear to sense; and finally stern resolve, a steely-eyed feeling of focus. I committed my mind to "get over" my fear and sadness and just attend to the details of what needed to be done. I ordered the necessary lab work, carefully administered the chemotherapy, watched closely for side effects, and intensely monitored my teacher/patient's progress. I went to the library and gathered all the research facts I could about his form of leukemia, the treatment, and the prognosis. I presented these papers and the "clinical case" to my team of fellow students, residents, and supervising doctors. On teaching rounds in the patient's room and beside his door, I discussed the technical details of the case with my attending and residents: just the facts, no feelings. I was careful not to spend much time talking with my patient. He was the sick one, I was the doctor. What was there to talk about, anyway?

Let me be clear: An intentional and temporary "just the facts" orientation can be a very useful stance to take at specific times. But *temporary* is the key—not a way of life, but a way of adapting, intentionally, in that moment, to a situation that requires that we act incisively and efficiently. To compartmentalize in this way is in itself a rigorous form of mental training. If you are being wheeled into the operating room, you want to encounter a confident, calm, task-oriented surgeon, not one who is upset or in tears. Even as parents in the face of a crisis, we need to focus our minds clearly on the problem at hand. Mindsight can allow us to ascertain that in such a situation, becoming upset or overidentifying with another person is not adaptive, and it can help us direct our attention to what needs to be done. But it can also help us to stay present with our own internal life and attuned to others, and to acknowledge our feeling-full mind, the "invisible" and richly subjective part of our lives.

When my internal medicine clerkship was finally over, a coveted "excellent" final grade was added to my record. And inside, I felt nothing. My heart was made of wood, driftwood, rotting on the shore, the waves lapping at the edge of a sea I no longer knew. The ether had returned.

THE MIND IS REAL, WHY NOT DEFINE IT?

Twenty-five years to the very week after I made the decision to drop out of medical school, I found myself back in the Ether Dome. The situation was a bit different. I had, after all, trained as a pediatrician and a psychiatrist, and I had been invited there, all those years later, to deliver a keynote address on the importance of emotions and stories in the development of health. My fifteen-year-old son, traveling with me on this lecture tour, was there in the audience, and I was filled with feelings I can hardly describe—of gratitude, of relief, of appreciation that so much had changed.

Over the last quarter century, science has opened a new window into the nature of our lives. We can state definitively that the mind, though not visible to the eye, is unequivocally "real." Medicine too has progressed since those days. Harvard Medical School has changed, and many programs today give at least some attention to notions such as empathy and stress reduction in student doctors and the importance of seeing the patient as a person. I would have had a much better experience becoming a doctor with such an internally focused, well-rounded curriculum.

The fields where I've spent my own working life, pediatrics and psychiatry and psychology, have each allowed me to dive deeply into the mental sea. After a research fellowship that enabled me to study attachment, memory, and narrative and to explore the ways the mind develops in families, I became an educator in the field of mental health. There in the Ether Dome, I was delivering a lecture about the nature of the mind and the importance of mindsight in health. I was also able to ask the audience a question—one I have now asked in talks to nearly eighty thousand mental health practitioners, from psychiatry to psychology, social work to occupational therapy.

As I began, I asked for a show of hands: "During your training, how many of you attended a course or lecture defining the mind or mental health?" The responses were easy to count. In numerous countries on four different continents, in lecture halls around our globe, the same statistic has emerged again and again: Only *2 to 5 percent* of people in these fields have ever been given even a single lecture that defined the very foundation of their specialty—the mind. For them, as for me during my own training, the focus has always been on mental illness, on categories of symptoms, and on treatment techniques designed to diminish disorders. Yes, the world is filled with mental pain, and we certainly have an important role to play in helping people to alleviate their suffering. But too often we are doing so without a clear vision of our goal, without exploring what a healthy mind might actually be. How strange. I would soon discover, in fact, that other research fields concerned with mental processes also seemed to have pursued their fascinating investigations without defining the mind that they were attempting to study.

The definition of mind I now use with my patients and students was the result of a remarkable collaboration. In 1992, I organized an interdepartmental group at UCLA to study the connections between the brain and the mind. I'd recruited forty scientists from a wide range of fields, including linguistics, computer science, genetics, mathematics, neuroscience, sociology, and of course developmental and experimental psychology. It was the beginning of the Decade of the Brain, and we were excited to tackle the tough questions about how the physical nature of the brain was somehow related to the subjective nature of the mind.

It quickly emerged, however, that each of the disciplines had its own way of seeing reality, and although we could easily agree that the brain was composed of a set of neurons encased in the skull and interconnected with the rest of the body, there was no shared view of the mind, and no common vocabulary for discussing it. A computer scientist referred to it as "an operating system." A neurobiologist said that "the mind is just the activity of the brain." An anthropologist spoke of "a shared social process passed across the generations." A psychologist said that "mind is our thoughts and feelings." And so it went, until I became worried that the tension from these differing perspectives in the group might lead to its dissolution. I had to create some acceptable working definition of the mind before we could address our fundamental seminar topic.

Here is the definition I ultimately offered to the group, a place to begin our explorations together: "The human mind is a relational and embodied process that regulates the flow of energy and information." That's it. Amazingly, every person in the group—from all the various fields involved—affirmed that this definition fitted with their own discipline's approach.

The mind is real and ignoring it does not make it go away. Defining the mind makes it possible for us, both in our daily life and in our many professional pursuits—from psychotherapy and medicine and education to policy formation and public advocacy—to share a common language about the internal nature of our lives.

To be sure that you and I share the same understanding, let's look in more detail at the elements of this working definition. I'll start at the end and work back towards the beginning.

THE MIND INVOLVES A FLOW OF ENERGY AND INFORMATION

Energy is the capacity to carry out an action—whether it is moving our limbs or thinking a thought. The various forms of energy are explored in physics and can be described in many different ways, but this essential "ability to do stuff" remains the same. We feel radiant energy when we sit in the sun, we use kinetic energy when we walk on the beach or go for a swim, we utilize neural energy when we think, when we talk, when we listen, when we read.

Information is anything that symbolizes something other than itself. These words you are reading, or words that you hear, are packets of information; the squiggles on the page are not the meaning of the words, and the words that you hear are just sound waves moving air molecules at certain frequencies. Conversely, a stone in itself is not information. The stone has data: We can weigh it and note its color, texture, and chemical composition. We can imagine the geological age when it was formed, and the many forces that have shaped it. But our minds are creating that information, and unless someone has carved a picture or a word on its surface, unless we think about its history or talk about it with one another, the stone is just a stone. On the other hand, the word *stone* is a packet of information. Even the idea of a stone can have meaning for you—but, again, that meaning is created by your mind, not by the stone itself.

Energy and information go hand in hand in the movement of our minds. We can have direct experience in the moment—for example, being aware of the sensations in our stomach when we're hungry, the flood of emotions when we're upset. We can also build on these energy-filled sensations and feelings by mapping them in the higher areas of our brain. We can "know" that stomach gurgling means we "should" eat, then look at the clock and tell ourselves to wait another half hour for lunch. We can interpret the meaning of an emotion—understanding an eruption of sadness in our heart as a response to the loss of a loved one, becoming aware of a resulting sense of isolation and loneliness—and then be motivated to do something about it, perhaps by seeking comfort from a friend. This is how our minds create information from the flow of energy and how information then leads to motivation and the exertion of energy in new and adaptive ways.

In chapter 1, I introduced the scientific term *representation* to convey this notion of information. Our ability to "represent" an emotional reaction to ourselves, to give it a name and a meaning, helps to lift us out of the immediacy of an experience so that we can respond to it effectively.

Knowing that our minds regulate the flow of both energy and information enables us to feel the reality of these two forms of mental experience—and then to act on them rather than get lost in them.

And what does it mean to say energy and information "flow"? Because they change across time, we can sense their movement from one moment to the next in a dynamic, fluid, moving process. But we don't just observe them. We can step into the river of time to change how those patterns unfold. The mind's regulation creates new patterns of energy and information flow, which we then continue to monitor and modify. This process is the essence of our subjective experience of life.

THE MIND IS A REGULATORY PROCESS: MONITORING AND MODIFYING

Consider the act of driving. To drive or "regulate" a car, you must both be aware of its motion and its position in space and also be able to influence how it moves. If you have your hands on the wheel but your eyes are shut (or focused on your text message), you can make the car move, but you're not driving it—because *driving* means regulating the car's movement, its flow, across time. If you have your eyes open but you're sitting in the backseat, you can monitor the movement of the car (and make comments, like one particular relative I know), but you can't actually modify its motion yourself. (No matter how hard you try. Sorry.)

If you are wondering what the "it" is that is being monitored and then modified by the mind, it is the flow across time of the two elements: energy and information. The mind observes information and energy flow and then shapes the characteristics, patterns, and direction of the flow.

Each of us has a unique mind: unique thoughts, feelings, perceptions, memories, beliefs, and attitudes, and a unique set of regulatory patterns. These patterns shape the flow of energy and information inside us, and we also share them with other minds. The powerful finding we'll explore throughout the rest of this book is that we can learn to shape these patterns, to alter our mind and also our brain, by first coming to see the mind clearly.

THE MIND IS EMBODIED AND RELATIONAL

We've now arrived at the beginning of the definition. When I say that the mind is *embodied,* I mean that the regulation of energy and information flow happens, in part, in the body. It occurs where we usually imagine our mental life taking place, in the circuits and synapses of the brain, inside the skull. But it also occurs throughout the body, in the distributed nervous system, which monitors and influences energy and information flowing through our heart and our intestines, and even shapes the activity of our immune system.

Finally, the mind is a *relational* process. Energy and information flow between and among people, and they are monitored and modified in this shared exchange. This is happening right now between you and me, through my writing and your reading. These packets of information—words on a page or spoken out loud—emerge from my mind and now enter yours. If we were sitting together in the same room, we would exchange all sorts of signals with each other, symbols we'd share in word form or in the nonverbal realm of eye contact, facial expression, tone of voice, posture, and gesture. Relationships are the way we share energy and information flow, and it is this sharing that shapes, in part, how the flow is regulated. Our minds are created within relationships—including the one we have with ourselves.

Offering this fundamental definition of a core aspect of the mind as "a process that regulates the flow of energy and information" was an important starting place for our interdisciplinary study group. This perspective created a foundation for us to explore other dimensions of our embodied and relational mind and what it means to be human.

INTERPERSONAL NEUROBIOLOGY

Our group continued to meet for four years, and since that time, an entire field has taken shape to build on this approach to the mind and mental health. Called "interpersonal neurobiology," it now has its own organizations, educational programs, and a professional library of more than a dozen textbooks. At the core of interpersonal neurobiology is our proposal that mindsight permits us to direct the flow of energy and information towards integration. And integration—which we'll be exploring in many of its real-world applications—is seen to be at the heart of well-being.

During this same time, new research into the mind-brain-body connection has demonstrated how our internal subjective states directly shape our physiological health. The negative impact of the stress hormone cortisol on our immune system's ability to fight infection and even cancer has been established. People exposed to emotional abuse as children have been found to be at higher risk of developing medical illnesses later in life, again possibly mediated by these stress effects on the body's defenses. And studies have shown that mindful awareness practices can improve the immune system's responsiveness.

I want to acknowledge, however, that bringing brain science into the day-to-day practice of psychotherapy, teaching, and medicine isn't for everyone, nor is everyone for it.

A senior clinician once said to me, "Dan, I've never seen a pre-frontal cortex in my life, so why should I think about one now?" Another confessed, "Thinking about the brain makes me feel stupid and incompetent, and I'm too set in my ways to make a change."

I've also been at professional meetings where clinical colleagues have told me that this approach is "bad." Since we don't know *everything* about the brain, why should therapists know *anything* about it? Another lecturer said that she thought it was "polluting the interpersonal space of therapy to bring in ideas from science about the brain." (These concerns I really didn't get. Why not build a framework, as we've done with interpersonal neurobiology, which is based solidly on science but deeply honors subjectivity and the importance of our interpersonal world?)

On the other hand, some neuroscientists are reluctant to embrace the notion of "mind" as more than "an outcome of the activity of the brain." The brain is a measurable entity with weight and volume, physical properties and a location. But where do we "find" the mind in physical space? How do we weigh it or assign a number to its features? At one meeting, a brain scientist declared, "We should never ask a question that cannot be quantified." Not to be outdone, a student of his raised the ante: "Well, we should never even think a thought that cannot be quantified." At this an anthropologist friend of mine turned a few shades of purple until he finally took a breath and expressed his stern disagreement. Many of us then gave an unquantifiable sigh of relief.

Of course sophisticated brain scans now allow us to do some of the quantifying: We can measure degrees of blood flow in the brain, the density of neural connections in a particular area, or the amplitude of electrical activity at a certain time. And as you can see in the

Minding the Brain sections that I've placed between the chapters, there is exciting new science tracing the brain activity correlated with some of our most intimate experiences. But much of the internal world is not quantifiable in absolute terms. How do we measure meaning? How do we assign a numerical value to a feeling or an intention? How can we quantify our sense of connection to one another, of "feeling felt," of being seen?

These discussions are not just academic, they are crucial to how we define our reality. Modern science is founded on measurement; it is a discipline based on statistics and numerical analysis that can be replicated and verified by objective observers. The subjective world of the mind, however, is observable primarily in qualitative terms, often based on unique first-person accounts by the person who actually has the mind in question. If you stick to the numbers game, the mind can easily disappear. When I'm in these challenging if sometimes frustrating academic discussions, I can't help remembering my experiences beneath the Ether Dome. Numerous esteemed faculty members in medicine and surgery seemed to live as if the mind did not exist. These were reasonable men and women, brilliant in their fields. How could something as real as the mind become so invisible to their—well, to their minds?

A REFINED VIEW OF MINDSIGHT

The mind is broader than the brain, revels in relationships, and is pregnant with possibilities. Yet this subjective core of our lived experience cannot be held in our hands or photographed with even the most majestic of machines. And the mind can easily be lost if we focus only on the domain of the physical. We can wipe away our tears and not leave even a trace of the mind that made meaning, that felt feelings, that enabled us to know we were alive and filled with pain, or joy.

When we perceive the mind, we are sensing something even more than our internal worlds or the inner lives of others: We have now refined our concept of mindsight beyond our initial description of it as a combination of insight and empathy. While this is an accessible and important place to start, it is just the beginning of a fuller story.

What mindsight does is enable us to sense and shape energy and information flow. That's the basic definition, the deeper truth, the fuller picture. With mindsight we gain perception and knowledge of the regulation (mind), sharing (relationships), and mediating neural mechanisms (brain) at the heart of our lives. "Our lives" means yours and mine. Mindsight takes away the superficial boundaries that separate us and enables us to see that we are each part of an interconnected flow, a wider whole.

By viewing mind, brain, and relationships as fundamentally three dimensions of one reality—of aspects of energy and information flow—we see our human experience with truly new eyes.

. . . .

Minding the Brain
Riding the Resonance Circuits

IT'S FOLK WISDOM THAT COUPLES in long and happy relationships look more and more alike as the years go by. Peer closely at those old photographs, and you'll see that the couples haven't actually grown similar noses or chins. Instead, they have reflected each other's expressions so frequently and so accurately that the hundreds of tiny muscle attachments to their skin have reshaped their faces to mirror their union. How this happens gives us a window on one of the most fascinating recent discoveries about the brain, and about how we come to "feel felt" by one another. Some of what I'll describe here is still speculative, but it can shed light on the most intimate ways we experience mindsight in our daily lives.

NEURONS THAT MIRROR OUR MINDS

In the mid-1990s, a group of Italian neuroscientists were studying the premotor area of a monkey's cortex. They were using implanted electrodes to monitor individual neurons, and when the monkey ate a peanut, a certain electrode fired. No surprise there—that's what they expected. But what happened next has changed the course of our insight into the mind. When the monkey simply *watched* one of the researchers eat a peanut, that same motor neuron fired. Even more startling: The researchers discovered that this happened only when the motion being observed was goal-directed. Somehow, the circuits they had discovered were activated only by an intentional act.

This mirror neuron system has since been identified in human beings and is now considered the root of empathy. Beginning from the perception of a basic behavioral intention, our more elaborated human prefrontal cortex enables us to map out the minds of others. Our brains use sensory information to create representations of others' minds, just as they use sensory input to create images of the physical world.

The key is that mirror neurons respond only to an act with intention, with a predictable sequence or sense of purpose. If I simply lift up my hand and wave it randomly, your mirror neurons will not respond. But if I carry out any act you can predict from experience, your mirror neurons will "work out" what I intend to do before I do it. So when I lift up my hand with a cup in it, you can predict at a synaptic level that I intend to drink from the cup. Not only that, the mirror neurons in the premotor area of your frontal cortex will get you ready to drink as well. We see an act and we ready ourselves to imitate it. At the simplest level, that's why we get thirsty when others drink, and why we yawn when others yawn. At the most complex level, mirror neurons help us understand the nature of culture and how our shared behaviors bind us together, mind to mind.

The internal maps created by mirror neurons are automatic—they do not require consciousness or effort. We are hardwired from birth to detect sequences and make maps in our brains of the

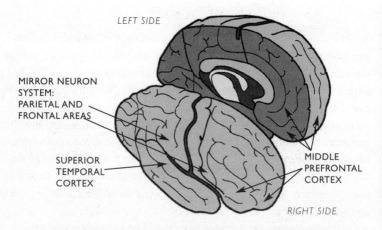

LEFT SIDE

MIRROR NEURON
SYSTEM:
PARIETAL AND
FRONTAL AREAS

SUPERIOR
TEMPORAL
CORTEX

MIDDLE
PREFRONTAL
CORTEX

RIGHT SIDE

The "resonance circuitry" includes the mirror neuron system (MNS), the superior temporal cortex, the insula cortex (not visible in this drawing, but linking these areas to the inner limbic region), and the middle prefrontal cortex.

internal state—the intentional stance—of other people. And this mirroring is "cross-modal"—it operates in all sensory channels, not just vision—so that a sound, a touch, a smell, can cue us to the internal state and intentions of another. By embedding the mind of another into our own firing patterns, our mirror neurons may provide the foundation for our mindsight maps.

Now let's take another step. Based on these sensory inputs, we can mirror not only the behavioral intentions of others, but also their emotional states. In other words, this is the way we not only imitate others' behaviors but actually come to resonate with their feelings—the internal mental flow of their minds. We sense not only what action is coming next, but also the emotional energy that underlies the behavior.

In developmental terms, if the behavioral patterns we see in our caregivers are straightforward, we can then map sequences with security, knowing what might happen next, embedding intentions of kindness and care, and so create in ourselves a mindsight lens that is focused and clear. If, on the other hand, we've had parents who are confusing and hard to "read," our own sequencing circuits may create distorted maps. So from our earliest days, the basic circuitry of mindsight can be laid down with a solid foundation, or created on shaky ground.

KNOWING ME, KNOWING YOU

I once organized an interdisciplinary think tank of researchers to explore how the mind might use the brain to perceive itself. One idea we discussed is that we make maps of intention using our cortically based mirror neurons and then transfer this information downward to our subcortical regions. A neural circuit called the insula seems to be the information superhighway between the mirror neurons and the limbic areas, which in turn send messages to the brainstem and the body proper. This is how we can come to resonate physiologically with others—how even our respiration, blood pressure, and heart rate can rise and fall in sync with another's internal state. These signals from our body, brainstem, and limbic areas then travel back up the insula to the middle prefrontal areas. I've come to call this set of circuits—from mirror neurons to subcortical regions, back up to the middle prefrontal areas—the "resonance circuits." This is the pathway that connects us to one another.

Notice what happens when you're at a party with friends. If you approach a group that is laughing, you'll probably find yourself smiling or chuckling even before you've heard the joke. Or perhaps you've gone to dinner with people who've suffered a recent loss. Without their saying anything, you may begin to sense a feeling of heaviness in your chest, a welling up in your throat, tears in your eyes. Scientists call this *emotional contagion*. The internal states of others—from joy and play to sadness and fear—directly affect our own state of mind. This contagion can even make us interpret unrelated events with a particular bias—so that, for example, after we've been around someone who is depressed we interpret someone else's seriousness as sadness. For therapists, it's crucial to keep this bias in mind. Otherwise a prior session may shape our internal state so much that we aren't open and receptive to the new person with whom we need to be resonating.

Our awareness of another person's state of mind depends on how well we know our own. The insula brings the resonating state within us upward into the middle prefrontal cortex, where we make a map of our internal world. So we feel others' feelings by actually feeling our own—we notice the belly fill with laughter at the party or with sadness at the funeral home. All of our subcortical data—our heart rate, breathing, and muscle tension, our limbic coloring of emotion—travels up the insula to inform the cortex of our state of mind. This is the brain reason that people who are more aware of their bodies have been found to be more empathic. The insula is the key: When we can sense our own internal state, the fundamental pathway for resonating with others is open as well.

The mind we first see in our development is the internal state of our caregiver. We coo and she smiles, we laugh and his face lights up. So we first know ourselves as reflected in the other. One of the most interesting ideas we discussed in our study group is that our resonance with others may actually precede our awareness of ourselves. Developmentally and evolutionarily, our modern self-awareness circuitry may be built upon the more ancient resonance circuits that root us in our social world.

How, then, do we discern who is "me" and who is "you"? The scientists in our group suggested that we may adjust the location and firing pattern of the prefrontal images to perceive our own mind. Increases in the registration of our own bodily sensations combined with a decrease in our mirror neuron response may help us know that these tears are mine, not yours—or that this anger is indeed from me, not from you. This may seem like a purely philosophical and theoretical question until you are in the midst of a marital conflict and find yourself arguing about who is the angry one, you or your spouse. And certainly, as a therapist, if I do not track the distinction between me and other, I can become flooded with my patients' feelings, lose my ability to help, and also burn out quickly.

When resonance literally becomes mirroring, when we confuse me with you, then objectivity is lost. Resonance requires that we remain differentiated—that we know who we are—while also becoming linked. We let our own internal states be influenced by, but not become identical with, those of the other person. It will take much more research to elucidate the exact way our mindsight maps make this distinction, but the basic issues are clear. The energy and information flow that we sense both in ourselves and in others rides the resonance circuits to enable mindsight.

As I consider the resonance circuits, two mind lessons stand out for me. One is that becoming open to our body's states—the feelings in our heart, the sensations in our belly, the rhythm of our breathing—is a powerful source of knowledge. The insula flow that brings up this information and energy colors our cortical awareness, shaping how we reason and make decisions. We cannot successfully ignore or suppress these subcortical springs. Becoming open to them is a gateway to clear mindsight.

The second lesson is that relationships are woven into the fabric of our interior world. We come to know our own minds through our interactions with others. Our mirror neuron perceptions, and the resonance they create, act quickly and often outside of awareness. Mindsight permits us to invite these fast and automatic sources of our mental life into the theater of consciousness. As we welcome the neural reality of our interconnected lives, we can gain new clarity about who we are, what shapes us, and how we in turn can shape our lives.

4

The Complexity Choir
Discovering the Harmony of Health

WHAT IS A HEALTHY MIND? Is it simply the absence of symptoms and dysfunctions, or is there something more to a life well lived? How can we embrace the diversity of behavior, temperament, values, and orientation across a wide range of cultures and still come up with a coherent definition of health? Just as some scientists are reluctant to define the mind, some people say that we shouldn't define mental health at all, because it is authoritarian to do so—we shouldn't tell others how to be healthy. But how do we account for the universal striving for happiness? How do we understand the cross-culturally recognizable ease of well-being? Positive psychology has offered an important corrective to the disease model by identifying the characteristics of happy people, such as gratitude, compassion, open-mindedness, and curiosity, but is there some unnamed quality that underlies all of these individual strengths?

Over the last twenty years, I've come to believe that integration is the key mechanism beneath both the absence of illness and the presence of well-being. Integration—the linkage of differentiated elements of a system—illuminates a direct pathway towards health. It's the way we avoid a life of dull, boring rigidity on the one hand, or explosive chaos on the other. In ways we'll explore in great depth in part 2, we can learn to detect when integration is absent or insufficient and develop effective strategies to promote differentiation and then linkage. The key to this transformation is cultivating the capacity for mindsight.

In new interventions based on the approach of interpersonal neurobiology, mindsight has helped many people shift the flow of energy and information in their lives towards integration. But why is integration such a powerful tool for transformation? My search for an answer to this question has led to some surprising and practical realizations.

THE CHOIR SINGS

These days, before I define mental well-being in my lectures, I often ask for volunteers to sing in a "complexity choir." Experienced singers usually break the ice and come bounding up to the front of the room, while others, initially more reticent, slowly find their way to join in. Whether my audience is parents or teachers, therapists or scientists, I know that the best way to help them grasp the power of integration is through immersion in direct experience.

My first request is that the newly assembled choir members all sing the same note at the same time, simply humming along in unison. Someone comes up with a mid-range pitch and they quickly settle into a uniform sound. After about half a minute, I hold up my hand to stop them and then make another request. This time I ask them to cover their ears so they can't hear one another, and then, at my signal, launch individually into whatever song with whatever words they'd like to sing. The audience usually laughs when the singers begin, but they quickly get restive, so I hold up my hand again.

Finally I ask the singers to choose a song most of them are likely to know and then to sing it together, harmonizing freely as the spirit moves them. This may be the ultimate pickup ensemble, but it's remarkable to hear what happens as a group of teachers or psychotherapists sail into "Oh! Susanna" or "Amazing Grace" or "Row-Row-Row Your Boat." (And it's fascinating to me that more than half the time, the group chooses "Amazing Grace"—which apparently is one of the most harmoniously balanced songs in the Western tradition.) Once the melody is established, individual voices begin to emerge, weaving their harmonies above and below, playing off one another, moving intuitively towards a crescendo before the final notes. Faces light up in choir and audience alike; we are all swept into the flow of the singers' energy and aliveness. At these times, people have said—and I've experienced this as well—there is a palpable sense of vitality that fills the room.

At that moment we are experiencing integration at its acoustic best. Each member of the choir has his or her unique voice, while at the same time they are linked together in a complex and harmonious whole. One is never quite certain where the choir will take the song, but the surprises simply highlight the pleasure of a familiar, shared melody. This balance between differentiated voices on the one hand and their linkage on the other is the embodiment of integration.

And what about the first two exercises? As you surely could predict, the single-note humming is unchanging, rigid—and after a while, dull and boring. The initial excitement and risk of volunteering gives way to the monotony of the task. The singers may be linked, but they cannot express their uniqueness, their individuality. When differentiation is blocked, integration cannot occur. Without the movement towards integration, the entire system moves away from complexity—away from harmony—and into rigidity.

On the other hand, when the singers close their ears and sing whatever they want, what emerges is cacophony, a chaotic outpouring of sound that often creates a sense of anxiety and distress in the listeners. Now there is no linkage—only differentiation. When integration is blocked in this way, we also move away from complexity, away from harmony. But this time we move towards chaos, not rigidity.

As the singers settle into their seats again, I sum up the point of the exercise: It is the middle way between chaos and rigidity—the flow of independent voices linked together in harmony—that maximizes both complexity and vitality. This is the essence of integration.

IN SEARCH OF INTEGRATION

When I first began to explore the idea of integration, it intuitively felt right that integration would be important to our individual and relational well-being. But I knew of no scientific explanation for why this might be the case.

Integration is mentioned, almost as an aside, in numerous disciplines, from the study of emotion and social functions to research into the brain itself. Yet none of these fields seem to give integration a central role, nor do they clarify why integration would be a good thing in life. Take for example the various scientific fields that study emotion. You might be surprised that there is no universal definition of emotion, even among emotion researchers. When I was reviewing the science of emotion for my first book, I discovered formulations like these: Emotion is a fundamental part of the person across the lifespan. Emotion connects body to brain. Emotion links one person to another. Each of these perspectives described an integrative process—yet integration itself was not discussed directly. Perhaps it was being an outsider to emotion research that helped me to see the common feature underlying their quite distinct definitions of what emotion is, what it does, and how it manifests itself in our lives across time.

What role could integration and emotion play in our definition of the mind as an embodied and relational process? Why do people use terms such as *emotional well-being* or *emotionally healthy* or *emotionally close* to label mentally healthy states? And what about such expressions as *emotional breakdown* or *emotionally upset*?

As a psychotherapist, I'd worked closely with many people in states of distress, states that to me seemed to be characterized as either rigidity or chaos—or both. Individuals might be stuck in depression or paralyzed by fear. They'd find themselves swept into manic rages or flooded with traumatic memories. Sometimes they'd fluctuate between these extremes, stuck in a whirlwind of energy and information, terrified by minds out of control.

But why rigidity or chaos? Why would dysfunction fall into these two categories, or some combination of the two? And why did these patterns keep recurring?

There was something about these states that seemed the antithesis of the harmony of a more integrated flow. Could these emotional shifts in our lives reflect changes in our states of integration? Perhaps the term *emotion* itself might be defined as "a shift in our state of integration." If so, emotion researchers—whatever their approach—might be able to agree that impairments to emotional well-being are movements of the mind away from integration. And perhaps—looking even deeper—integration might be the principle underlying health at all levels of our experience, from the microcosm of our inner world to our interpersonal relationships and life in our communities.

A HEALTHY MIND: COMPLEXITY AND SELF-ORGANIZATION

Diving again into the scientific literature, I finally came across an unlikely discipline that could be relevant to our exploration of the mind: a branch of mathematics that focuses on complex systems. Here was a plausible scientific foundation for the benefits of integration—a reason integration is a good thing in our lives.

In brief, complexity theory examines systems that are capable of becoming chaotic and are open to receiving input from outside themselves. Thinking in systems terms requires that we focus on the relationships among the elements that interact to compose the "system." One classic example of a complex system is a cloud—a collection of water molecules capable of random distribution (it can be chaotic), and which receives light and energy such as wind and heat from outside itself (it is open). Complexity theory explores the natural movements of this open and chaos-susceptible system across time—explaining, for example, why clouds emerge, change shape, and dissipate. It seemed to me that human lives also meet these criteria—we are open systems capable of chaotic behavior—so I read on.

A complex system is said to regulate its own emergence. This means that the system itself has certain properties that determine how it unfolds over time. This self-organizational process, the way the system shapes its own unfolding, is built from the mathematics of complex systems. There is no programmer, no program, no outside force governing how the system will flow across time. Self-organization emerges from the interactions among the basic elements that comprise the system. Again, if self-organization applies to clouds, it likely applies to other open systems capable of chaos. We are certainly capable—sometimes too much so—of becoming chaotic. And we are quite open to influences outside of ourselves—from people we meet, experiences we have in the world, books we read. If these ideas were relevant and true, then perhaps this was an argument for the idea that we too are capable of self-organization. It seemed to me that our triangle of well-being, the system of mind, brain, and relationships, might be more fully understood in these terms, and we might apply the principles of complexity and integration to creating health across each of these three aspects of our lives.

THE RIVER OF INTEGRATION: RIGIDITY OR CHAOS VERSUS HARMONY AND FLEXIBILITY

A system that moves towards complexity is the most stable and adaptive. Reading this for the first time in the literature on the mathematics of complex systems, I thought, What a clear definition of well-being! I jumped up and pulled off my shelf the 886-page psychiatrists' bible, the *Diagnostic and Statistical Manual of Mental Disorders*. I decided to open it at random to any page. There it was: Wherever I put my finger, on whatever symptom of whatever dysfunction, there was an example of chaos, rigidity, or both. Could it be that mental health was indeed a function of integration? When our minds move away from integration, away from harmony, are we then prone to live in chaos and/or rigidity?

I began to try out this hypothesis on my colleagues and students, and even though some of them found it rather new and strange, it seemed to fit their experience as clinicians. Then I started to apply it to my own work with patients, exploring ways to promote integration as a framework for helping them move from illness to wellness. Just like that, fresh approaches to treatment began to emerge, some of them startlingly effective. This notion of the central role of integration was and remains an amazing organizing perspective that has enabled me and now my colleagues to promote well-being in powerful new ways.

I am an acronym lover, always looking for ways to make clusters of related items stick in my mind—and to make them easier to teach. One day in a seminar, I asked my students for suggestions about how we could remember the flow of an integrated system. "Oh Dan, that's easy," a young woman replied. "Just remember Saks Fifth Avenue: Stable, Flexible, and Adaptive." I thought for a moment and then pointed to my clothes. There was the evidence that this mnemonic probably would not work for me.

I also wanted to capture the sense of vitality and energy that emerges from the complexity choir at its harmonious best. Later that day, an acronym came to me: SAFE, as in Stable, Adaptive, Flexible, and Energized. And then a few weeks later, after reading more into the mathematics of something called "coherence," I realized that coherence was a fifth essential characteristic of integration, which fit beautifully with my own area of research, which had found that "coherent narratives"—the way we make sense of our lives and free ourselves from the prisons of the past—are an important predictor of relational health (as we'll explore in part 2).

Now the qualities of an integrated flow spelled a universally memorable word: FACES, for Flexible, Adaptive, Coherent, Energized, and Stable. We can say that any healthy complex system has a FACES flow. In other words, when the self-organizational movement of the system is maximizing complexity, it attains a harmonious flow that is at once flexible, adaptive, coherent, energized, and stable. This is the feeling you get from our amazing and graceful complexity choir.

I like to imagine the FACES flow as a river. The central channel of the river is the ever-changing flow of integration and harmony. One boundary of this flow is chaos. The other boundary is rigidity. These are the two banks of the river of integration.

Sometimes we move towards the bank of rigidity—we feel stuck. Other days we lean towards chaos—life feels unpredictable and out

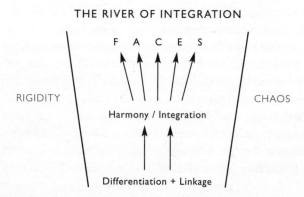

THE RIVER OF INTEGRATION

F A C E S

RIGIDITY   CHAOS

Harmony / Integration

Differentiation + Linkage

of control. But in general, when we are well and at ease, we move along this winding path of harmony, the integrated flow of a flexible system. We sense the familiar but are not trapped by it. We live at the threshold of the unknown and have the courage to move into new and uncharted waters. This is living a life as it unfolds, moment by moment, in a flowing journey between rigidity and chaos. This is the FACES flow. An old, dear and now-departed friend, the poet, philosopher, and all around wonderfully wise John O'Donohue, captured the essence of this emergent flow when he said that he'd love to live like a river, carried by the surprise of his own unfolding.

THE EIGHT DOMAINS OF INTEGRATION

In my practice of psychotherapy, eight domains of integration have emerged as keys to personal transformation and well-being. These domains do not necessarily develop in a linear fashion, and in the chapters of part 2, you'll see that they sometimes emerge in combination. How we experience a "sense of self"—a feeling of who we are over time and of the patterns of energy and information that unfold in our inner lives—will be directly shaped by the degree of integration in these domains.

Each of us has a different mind, so if you've had enough of conceptual overviews at this point, feel free to skip right to the stories in part 2. If not, here is a brief map of the domains that each of those stories will illustrate and expand on.

THE INTEGRATION OF CONSCIOUSNESS

How we focus our attention is the key to promoting integrative changes in the brain. With the integration of consciousness, we actually build the skills to stabilize attention so that we can harness the power of awareness to create choice and change. This is why the integration of consciousness is the foundation for the other domains. Creating what I'll call a "hub of awareness" enables us to acknowledge troubling states without being taken over by them, and to see things as they are, rather than being constrained by our expectations of how they "should be." It also opens us to the full range of our perceptions—to information from the external world, from our bodily states, from relationships, and from the mind itself.

We'll explore how the integration of consciousness can help regulate mood and emotion, calm internal storms, and cultivate a more flexible and stable mind. The lessons learned from stabilizing the mind through the integration of consciousness will be applied to all of the challenges of real life we'll encounter throughout the book.

HORIZONTAL INTEGRATION

For millions of years, our left brain and right brain have had separate but complementary functions. The right side develops early and is the realm of imagery, holistic thinking, nonverbal language, autobiographical memory, and a host of other processes. Our left brain develops later in life and is responsible for logic, spoken and written language, linearity, lists, and literal thinking. If the linkage between the sides is blocked, one side may dominate, and we can lose the creativity, richness, and complexity that results from both sides working together. Harnessing the power of neuroplasticity to integrate the brain can give us a newly coherent sense of our life story and deeper insights into the nonverbal world of ourselves and others.

In chapter 6 we'll meet an individual who for nearly a century has lived a life leaning to the left. With specific strategies to help develop his right hemisphere, this man was able to come to experience the energy and vitality of a newly integrated life.

VERTICAL INTEGRATION

Our nervous system is vertically distributed, ascending from the body proper through the brainstem and limbic areas and finally arriving at the cortex. From head to toe and back again, vertical integration links these differentiated areas into a functional whole. Vertical integration can be impaired in response to trauma or in adaptation to living in an emotional desert. In this cut-off state, we ignore what our senses and bodily sensations are telling us and live a life of flattened feelings and perceptions. Bringing our sensations into awareness enables intuition to blossom and sometimes can offer lifesaving information.

Even after years of living only "above her shoulders," the anxious and "disconnected" woman we meet in chapter 7 was able to find relief by learning to become open to the sensations of her body. Beyond being able to live with more vitality and gusto, she also learned to tap into the deep source of intuition and wisdom that become available through vertical integration.

MEMORY INTEGRATION

We process and encode our experiences in layers of memory. The first layer, implicit memory, begins in the womb and predominates throughout our early years. From our emotions, perceptions, actions, and bodily sensations, we create mental models that shape our expectations about the way the world works. All of this occurs without effort or intention, and our implicit mental models can continue to shape how we act without our awareness. The puzzle pieces of implicit memory are later assembled into explicit memories—the factual and autobiographical information of which we are aware. The more we can shine the light of mindsight on the free-floating puzzle pieces of the past—the implicit memories—and allow them to become explicit, the more we can free ourselves to live fully in the present and have new choices about how we live our lives.

In part 2 we'll meet many people whose impaired integration of memory prevented them from achieving coherence in their lives. Sometimes an overwhelming event, called trauma, can cause a person to remain in this unintegrated state, resulting in a tendency towards either rigid states of avoidance or intrusive states of chaos. Focusing mindsight's lens on these layers of memory can be an essential step in the resolution of trauma and the integration of the brain's memory functions.

NARRATIVE INTEGRATION

We make sense of our lives by creating stories that weave our left hemisphere's narrator function with the autobiographical memory storage of our right hemisphere. Research has revealed that the best predictor of the security of our children's attachment to us is our ability to narrate the story of our own childhood in a coherent fashion. By detecting blockages to narrative integration and then doing the necessary work to overcome them, we can free ourselves and ultimately our children from the cross-generational patterns we want to avoid creating.

We'll review how research findings and clinical experience with attachment illuminate the varied forms of narratives we have and how strategies to promote integration can move these cohesive but constrictive life stories towards coherence and flexibility. When we are able to "make sense" of our lives in a deep, integrative manner, what emerges is a coherent narrative of our lives.

STATE INTEGRATION

Each of us experiences distinct states of being that embody our fundamental drives and needs: closeness and solitude, autonomy and independence, caregiving and mastery, among others. These states may also conflict with one another—sometimes painfully and confusingly. Mindsight permits us to embrace these states as healthy dimensions of a layered life instead of as parts of ourselves that we need to reject or suppress to try to achieve inner stability.

With state integration, we can move beyond past patterns of adaptation and denial to become open to our needs and able to meet them in different ways at different times. We'll explore how facing some of our many states is an essential first step in differentiating our "multiple selves." The key to integration is then to embrace these distinctions rather than to attempt to deny their existence. The power of state integration to release us from the patterns of shame and terror that can paralyze us will be revealed in one man's journey of transformation.

INTERPERSONAL INTEGRATION

This is the "we" of well-being. At best, our resonance circuits enable us to feel the internal world within others, while they in turn weave us into their inner world and carry us with them even when we are not together. Mindsight can help us to see how past adaptations are restricting current relationships and then allows us to open ourselves safely to others. Then we can connect more intimately in relationships while still retaining our own sense of identity and freedom. We can love and be loved without giving up our selves.

We'll see how couples lost in confusion and misunderstanding, struggling with rigid patterns of defense or prone to chaotic outbursts of disillusionment, can be taught how to detect their own brainstem-driven states of reactivity and move their nervous systems towards the receptive state necessary for true and lasting connection. Knowing how the past has shaped the present through synaptic changes early in life, couples can then ease the hostility that often surrounds their dysfunctional relationships. We'll see how people can use mindsight to guide their way back to a life of passion and compassion as they promote integration within and between themselves.

TEMPORAL INTEGRATION

Uncertainty, impermanence, mortality: These are the profound challenges presented to us by the prefrontal cortex, which gives us both our sense of time and our ability, apparently unique among animals, to foresee that death will undo us and those we love. Obsessive-compulsive disorder reveals how our hardwired survival drive seeks control—sometimes to the point of paralyzing and terrorizing us. Temporal integration enables us to live with more ease and to find comforting connections in the face of uncertainty.

We'll explore how even young people afflicted with anxiety about death and uncertainty—manifested in obsessions or in existential dread—can find a way to integrate these temporal prefrontal issues into their lives and grow stronger because of them.

MINDSIGHT AND FREEDOM

Within each of us is an inherent drive towards health—a push towards integration. But life happens, and we may sometimes find that integration is blocked. This blockage can come from impairments to *linkage,* as in unresolved trauma. Blockage can also arise from impairments to *differentiation,* whether as a fallout from childhood neglect or as a result of various learning disabilities and developmental difficulties. Or both differentiation and linkage may be impaired.

Mindsight is the skill that can lead us back to integration. Michelangelo is supposed to have said that his great task as a sculptor was to liberate the figure from the stone. Just so, our task is to find the impediments to the eight domains of integration and liberate the mind's natural drive to heal—to integrate mind, brain, and relationships in the triangle of well-being.

As these eight domains of integration are created and developed, a new dimension of interconnection, which I have come to describe as "transpiration," or "breathing across," seems to emerge. I have seen this happen time and again in patients who have done the work of mindsight. Their identity expands; they become aware that they are part of a much larger whole. In various research explorations of happiness and wisdom, this sense of interconnection seems to be at the heart of living a life of meaning and purpose. This is the promise of mindsight and integration.

PART II

· · · · · · · ·

The Power to Change:
Mindsight in Action

5

A Roller-Coaster Mind
Strengthening the Hub of Awareness

JONATHON WAS IN HIGH SCHOOL, just turning sixteen, when I first met him. He shuffled into the room, his jeans precariously low on his hips and his lanky blond hair covering his eyes, and told me that he had felt "bad" and "down" over the last two months, with bouts of crying that seemed to come out of nowhere. In response to my questions, he reported that he had a close group of friends at school, and though his classes were demanding, nothing in particular had changed either academically or socially that would account for his dark moods. Life at home had been "fine," he said with a bland, almost dismissive tone. His older sister and younger brother were giving him their "usual grief" and his parents were being their "typical irritating selves"—nothing out of the ordinary for a sixteen-year-old boy, it seemed.

But something was very wrong. Jonathon's bouts of tears and feeling down were accompanied by bursts of rage he could not control. Ordinary incidents, such as his sister arriving late for the school run or his brother using his guitar without permission, could lead to screaming anger. This lowering of his reactivity threshold concerned not only his parents and me, but Jonathon as well. He told me sheepishly that these explosions of rage, while not new, were getting worse. Now they were beginning to scare him. Similar episodes had occurred several times since he began middle school at thirteen, but Jonathon's parents had attributed these times of emotional instability to adolescence—"just a part of being a teenager"—and hadn't made much of them until now. When he told them he sometimes felt that he couldn't go on living, they brought him to me for an evaluation.

AN UNRELIABLE MIND

What shapes the currents of our sea inside? When we hit rough waters, is there anything we can do to calm the storm? In this chapter I will explore how we can use focused, conscious attention first to sense, then to alter, the wild flow of energy and information that can plague our lives. This focused attention permits us to use awareness to create choice and change. This is the domain of the integration of consciousness.

The term *mood* refers to the overall tone of our internal state. We express this emotional baseline through our affect, the external signals that reveal our feelings, and by way of our actions and reactions. Simply sitting with Jonathon in my office, I could begin to pick up his feelings of despair and depletion. As he readily admitted, his down mood also included tearfulness, irritability, difficulty sleeping, and decreased appetite. He also admitted that his feelings of hopelessness and despair were sometimes accompanied by suicidal thoughts, but I was able to determine that he had made no attempts and had no plans, at least at the moment, to hurt himself.

In a psychiatric textbook this cluster of symptoms would point to a probable diagnosis of major depression, but as a clinician I wanted to keep an open mind about other potentially relevant issues. Jonathon's family history included both drug addiction in an uncle on his mother's side and manic-depressive illness (also known as bipolar disorder) in a grandfather on his father's side. This made me cautious about a premature diagnosis of only depression.

Because of the family history of drug abuse, Jonathon's family had already been screening him regularly for drug use. The tests were consistently negative, and Jonathon himself asked, "Why should I take things that would make me more up and down? They just mess me up even more than I already am." I was struck by his insight, and I believed him.

The abrupt explosions that took him down the low road might signal the irritability that is a hallmark of major depression, especially in children. But they could also be a symptom of bipolar illness, which often runs in families and can emerge in childhood and adolescence. In its initial presentation, bipolarity can be indistinguishable from what is called "unipolar" depression, in which the mood disturbance moves in one direction only: towards down, depressed states. In bipolar disorder, however, these depressed states alternate with the "up" (or, more accurately, "activated") state of mania. Adults and adolescents with mania can experience rapid thinking, an inflated sense of self-importance and power, decreased need for sleep, increased appetite (for both food and sex), excessive spending, and irrational behaviors.

Making the distinction between unipolar and bipolar mood disorders is crucial for proper treatment, so I often get a second opinion from a colleague regarding this diagnosis. In Jonathon's case we also got a third. Both confirmed my concern that Jonathon's mood disturbance might be emerging bipolar disorder.

Described in brain terms, bipolar disorder is a condition characterized by severe "dysregulation," meaning difficulty in maintaining equilibrium in the face of daily life. The sense one gets as a clinician is that there is a problem with the coordination and balance of the brain's mood-regulating circuits. As you've seen in the first Minding the Brain section, our subcortical regions influence our emotional states, altering our moods, coloring our feelings, and shaping our motivations and behaviors. The prefrontal cortex, sitting atop the subcortical areas, regulates how we bring these emotional states into equilibrium.

The regulatory circuits of the brain can malfunction for a number of reasons, some of them related to genetics or the constitutional (not learned) aspects of temperament. One current theory is that people with bipolar disorder may have a structural difference in the way their regulatory prefrontal circuits connect with the lower, emotion-creating and mood-shaping limbic areas. This anatomical difference, perhaps established by way of genetics, infection, or exposure to neurotoxins, may lead to the unbridled firing of lower limbic areas. When revved up, these subcortical circuits shape the rapid thinking, heightened appetites, and overall driven quality of the manic state. While mania may appear attractive and pleasurable to an observer, and the person experiencing it may indeed enjoy some periods of euphoria, he is also likely to have periods of agitation, irritability, and restlessness that feel out of control and desperate. And when the dysfunction in the subcortical circuits goes in the opposite direction, thought slows down, mood becomes depressed, the vital functions of sleep and appetite are disturbed, and the person may withdraw almost entirely from social contact. When impaired prefrontal regulation results in failure to bring these two extremes of the emotional continuum into equilibrium, both the manic and the depressive states can be experienced as extremely distressing.

The standard treatment for bipolar disorder is medication, which has clear benefits for many patients. However, the side effects of the medications used for bipolar disorder—called "mood-stabilizing agents"—are much more significant than those of the antidepressants used for unipolar depression. These risks present a serious set of considerations for child psychiatrists, making us hesitant to rush to the more long-term medications called for by a bipolar diagnosis. Furthermore, if someone with undiagnosed bipolar disorder presents first with depression and is given an antidepressant medication, that clinical intervention can actually trigger the onset of manic episodes. It may also make the individual prone to an intense form of the disorder with rapid cycling between mania and depression and sometimes the emergence of a "mixed state" of both extremes at the same time.

Taking all of these concerns into account, I asked Jonathon's parents to come in with him and we discussed the issues openly, including the role of medications in the treatment of serious psychiatric disturbances. Many clinicians focus primarily on the concept of "chemical imbalance," and how various neurotransmitters, such as serotonin or noradrenaline, take you "up" or "down" as their levels rise or fall. However, I actually find that a deeper discussion of emotional regulation in the brain gives patients a larger view of the problem—and what we can do about it. I introduced Jonathon and his family to the hand model of the brain and described the prefrontal region's crucial role. We didn't know *why* these circuits were not working optimally in Jonathon, I told them. We just knew that his severe mental storms likely correlated with such prefrontal dysfunction.

"What can be done to help those circuits work well?" Jonathon's mother asked astutely. One theory about depression, I said, is that the brain's ability to change in response to experience is shut down. (In terms of our river of integration we can see this as rigidity.) Antidepressants such as the familiar serotonin medications, the selective serotonin reuptake inhibitors, or SSRIs, and mood stabilizers such as lithium seem to help reignite neuroplasticity. They help change the brain both by altering the way neurotransmitters function and by enhancing the brain's ability to learn from experience—as in therapy. Medications and psychotherapy combined often make an excellent treatment strategy for major mood disorders. Even psychotherapy alone has been shown to change the way the brain functions. In fact, I told them, some recent findings have revealed that chronically relapsing episodes of depression, like the ones that Jonathon might be experiencing, may actually be prevented by a form of therapy based on an ancient technique called "mindfulness."

A MINDFUL APPROACH TO CHANGING THE MIND

At the time Jonathon came to me, I was in the midst of writing a book that reviewed the existing neuroscience research on mindfulness. Being mindful, having mindful awareness, is often defined as a way of intentionally paying attention to the present moment without being swept up by judgments. Practiced in the East and the West, in ancient times and in modern societies, mindful awareness techniques help people move towards well-being by training the mind to focus on moment-to-moment experience. People sometimes hear the word *mindfulness* and think "religion." But the reality is that focusing our attention in this way is a biological process that promotes health—a form of brain hygiene—not a religion. Various religions may encourage this health-promoting practice, but learning the skill of mindful awareness is simply a way of cultivating what we have defined as the integration of consciousness.

As I'd told Jonathon and his parents, research had clearly demonstrated that mindfulness-based therapy could help prevent relapse in people with chronic depression. I had found no comparable published research on using mindfulness for those with bipolar disorder.

However, I had reason to be cautiously optimistic. Controlled studies had shown that mindfulness could be a potent part of successful treatment for many conditions, including anxiety, drug addiction (both treatment and relapse prevention), and borderline personality disorder, whose hallmark is chronic dysregulation.

In fact, one of the first studies to reveal that psychotherapy could actually change the brain—a study of obsessive-compulsive disorder done at UCLA—used mindfulness as a component of the treatment. In addition, in our own pilot study at the Mindful Awareness Research Center, also at UCLA, we found that mindfulness training was highly effective for adults and teens who had trouble paying attention at work or school.

Would Jonathon's mood disorder respond to such an intervention? The family's cooperative stance, coupled with their concerns about medication's side effects, made me think it was worth trying. I sought Jonathon's and his parents' informed consent, keeping in mind his recent suicidal thoughts and the serious risks of untreated depression, whether unipolar or bipolar. We elected to do a trial of mindfulness training, agreeing that if it did not begin to work within a few weeks' time to reduce his suffering and stabilize his mood, we would turn to the next phase of treatment, which would probably include medication.

FOCUSING ATTENTION, CHANGING THE BRAIN

As I'd explained to Jonathon and his parents, the brain changes physically in response to experience, and new mental skills can be acquired with intentional effort, with focused awareness and concentration. Experience activates neural firing, which in turn leads to the production of proteins that enable new connections to be made among neurons, in the process called neuroplasticity. Neuroplasticity is possible throughout the lifespan, not just in childhood. Besides focused attention, other factors that enhance neuroplasticity include aerobic exercise, novelty, and emotional arousal.

Aerobic exercise seems to benefit not only our cardiovascular and musculoskeletal systems, but our nervous system as well. We learn more effectively when we are physically active. Novelty, or exposing ourselves to new ideas and experiences, promotes the growth of new connections among existing neurons and seems to stimulate the growth of myelin, the fatty sheath that speeds nerve transmissions. Novelty can even stimulate the growth of new neurons—a finding that took a long time to win acceptance in the scientific community.

Where we focus our attention channels our cognitive resources, directly activating neural firing in associated areas of the brain. For example, research has also shown that in animals rewarded for noticing sounds, the brain's auditory centers expanded greatly, while in those rewarded for attending to sights, the visual areas grew. The implication is that neuroplasticity is activated by attention itself, not only by sensory input. Emotional arousal may also be a factor in the activation that occurs when animals are rewarded for noticing sounds or sights, and the same factor may be involved in activating neuroplasticity when we participate in an activity that is important or meaningful to us. But when we are not engaged emotionally, the experience is less "memorable" and the structure of the brain is less likely to change.

Other evidence of brain reshaping as a result of focusing comes from brain scans of violinists. The scans show dramatic growth and expansion in regions of the cortex that represent the left hand, which must finger the strings precisely, often at very high speed. Other studies have shown that the hippocampus, which is vital for spatial memory, is enlarged in taxi drivers.

A MINDFUL BRAIN

The ability to focus the mind is what I wanted Jonathon to acquire through mindfulness training. But what exactly does mindful awareness training stimulate? And why would mindfulness, as research has shown, help with such a wide variety of difficulties, from mood to attention, addiction to personality disorders? Finally, could mindfulness training help Jonathon with his serious problem with dysregulation?

In summary, here is what modern clinical research, 2,500 years of contemplative practice, recent neuroscience investigations, and my own experience all suggest: Mindfulness is a form of mental activity that trains the mind to become aware of awareness itself and to pay attention to one's own intention. As researchers have defined it, mindfulness requires paying attention to the present moment from a stance that is nonjudgmental and nonreactive. It teaches self-observation; practitioners are able to describe with words the internal seascape of the mind. At the heart of this process, I believe, is a form of internal "tuning in" to oneself that enables people to become "their own best friend." And just as our attunement to our children promotes a healthy, secure attachment, tuning in to the self also promotes a foundation for resilience and flexibility.

The way that mindfulness seemed to overlap with the processes of secure attachment and with the key functions of the prefrontal region that I discussed in part 1 made a powerful impression on me. It seemed that the act of attunement—internal in mindfulness, or interpersonal in attachment—might lead to the healthy growth of middle prefrontal fibers. Shortly after I had this realization, I read a report of ongoing research that showed that the middle prefrontal regions were indeed thicker in mindfulness practitioners.

So this is the hypothesis that led me to offer mindfulness training to Jonathon: that the practice would help the parts of his brain that regulate mood to grow and strengthen, stabilizing his mind and enabling him to achieve emotional equilibrium and resilience. It is not that I believed he had a history of an insecure attachment, but rather that mindful awareness might directly stimulate the growth of the cluster of neurons called the resonance circuits, which I discussed in the third Minding the Brain segment. These neural circuits, which include the middle prefrontal areas, enable us to resonate with others and to regulate ourselves. It is here that we can see the connection between attunement and regulation: internal and interpersonal forms of attunement each lead to the growth of the regulatory circuits of the brain. When we have attunement—either interpersonally or internally—we become more balanced and regulated. Helping Jonathon achieve this form of internal attunement with mindfulness practice was our goal. This would take focus, time, and careful monitoring to be sure his underlying dysregulation did not worsen or endanger him or others.

THE ADOLESCENT BRAIN AND THE PREFRONTAL CORTEX

Jonathon was eager to find a way to ease his suffering. Normal adolescence is hard enough: negotiating the changes in one's body, the emerging and sometimes overwhelming feelings of sexuality; changes in self-identity and relationships; academic demands; uncertainties about the future; and the stresses in family life in anticipation of leaving home. The adolescent brain itself is in flux. The prefrontal regions, including the middle areas, do not mature fully until well into the mid-twenties. Not only is the brain exposed to dramatic hormonal changes, but it undergoes genetically programmed "neural pruning sprees"—the removal of neural connections to hone down the various circuits, preserving those that are used and discarding the unused, so that the brain becomes more specialized and efficient. The normal remodeling of the brain is intensified by stress, and it can unmask or create problems during this vulnerable period. This makes the nine middle prefrontal functions—from fear modulation to empathy and moral awareness—somewhat unpredictable, so that emotional self-regulation can be challenging for any teenager.

Jonathon's mood dysregulation went well beyond normal adolescent turmoil. Most adolescents do not get to the point of suicidal thinking, or to a place where their unpredictable moods create significant chaos in their lives. These eruptive and painful periods had created self-doubt in Jonathon. He felt he could no longer depend on his own mind, that his mind was betraying him.

It seemed to me that becoming "his own best friend" was exactly what Jonathon needed. If we could help him grow the integrative fibers of his middle prefrontal cortex, he might be able to achieve more of the FACES flow I discussed in chapter 4, so that he could find a more harmonious path between the banks of rigidity and chaos. Integration of consciousness might help stabilize his mind.

I explained all of this to Jonathon, and reminded him that with regular exercise, a good diet, and sleep, he could set the foundation for promoting neuroplasticity. Jonathon and I made a verbal agreement that he would follow this "prescription" for health. It's amazing how often these basics of brain health are ignored. Exercise is an underrated treatment—and now we know that aerobics not only releases the endorphins that can combat a down mood but also promotes the growth of the brain. Eating regularly and well, balancing the various food groups, and avoiding excessive sugar and stimulants can help to reduce mood swings. And sleep, though in short supply and difficult at times to initiate for Jonathon, is a healer that can be approached in a systematic way. Sleep hygiene includes setting up a calming routine before bed. Minimizing caffeine or other stimulants once evening approaches, if not before; shutting off digital stimulation an hour or two before sleeping; and quiet activities such as taking a bath, listening to soothing music, or reading a book can all help the body as well as the mind to settle. With these brain hygiene basics in our contract, we could move into our specific efforts to promote integration.

Now it was time to use the focus of Jonathon's mind to change his brain. We began a series of skill-training sessions to help him develop mindful awareness. The idea was that the techniques I taught him would create a temporary state of brain activation each time they were repeated. Induced regularly, these temporary states would become long-term, enduring traits. With practice, a mindful state becomes a mindful trait.

THE WHEEL OF AWARENESS: RIM, SPOKES, AND HUB

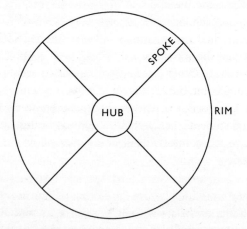

This is the basic diagram I drew for Jonathon to help him visualize how we can focus our attention.

A picture in my own mind helped me make sense of the techniques I'm about to describe to you. I call it the mind's "wheel of awareness." I drew it for Jonathon as we started our work together. Picture a bicycle wheel, with the hub at the center, and spokes radiating to the outer rim. The rim represents anything we can pay attention to, such as our thoughts and feelings, or our perceptions of the outside world, or the sensations from the body. The hub represents the inner place of the mind from which we become aware. The spokes represent how we direct our attention to a particular part of the rim. Our awareness resides in the hub and we focus on the various objects of our attention as points on the rim. The hub can be seen as a visual metaphor for our prefrontal cortex. To experience this directly, let's turn to the first exercise I offered to Jonathon.

A MINDFUL AWARENESS EXERCISE: FOCUSING ON THE BREATH

Over thousands of years of human history, from East to West, virtually all cultures have developed some form of practice that harnesses the power of mindfulness to cultivate well-being. These include body- and energy-centered practices such as yoga, tai chi, and qigong; devotional practices such as centering prayer or chanting; and various forms of sitting and walking meditation that were first introduced into the West by Buddhist practitioners.

I elected to teach Jonathon a practice called "insight meditation," both because I had learned it myself from experienced teachers and because it had the most research backing up its potential to help develop the brain. Other techniques might have been just as reasonable a starting point, but I felt most comfortable with this one.

Here is a transcript of the meditation exercise that I teach my patients and students. Feel free to read through this, and then try it out if you're in a comfortable place that will allow you to dive into the sea inside.

It's helpful to be able to become aware of your own mind. That can be a very useful awareness to have. Yet not much happens in school or in our family life that lets us come to know ourselves. So we are going to spend a couple of minutes now doing just that.

Let yourself get settled. It's good to sit with your back straight if you can, feet planted flat on the floor, legs uncrossed. If you need to lie flat on the floor that's okay, too. And with your eyes open at first, just try this. Try letting your attention go to the center of the room. And now just notice your attention as you let it go to the far wall. And now follow your attention as it comes back to the middle of the room and then bring it up close as if you were holding a book at reading distance. Notice how your attention can go to very different places.

Now let your attention go inward. You might let your eyes close at this point. Get a sense inside yourself of your body in space where you're sitting in the room. And now let yourself just become aware of the sounds around you. That sense of sound can fill your awareness. (Pause for some moments.)

Let your awareness now find the breath wherever you feel it most prominently—whether it's at the level of your nostrils, the air going in and out, or the level of your chest as it goes up and down, or the level of your abdomen going inward and outward. Perhaps you'll even just notice your whole body breathing. Wherever it comes naturally, just let your awareness ride the wave of your in-breath, and then your out-breath. (Pause.)

When you come to notice, as often happens, that your mind may have wandered and become lost in a thought or a memory, a feeling, a worry, when you notice that, just take note of it and gently, lovingly, return your awareness towards the breath—wherever you feel it—and follow that wave of the in-breath, and the out-breath. (Pause.)

As you follow your breath, I'm going to tell you an ancient story that's been passed through the generations.

The mind is like the ocean. And deep in the ocean, beneath the surface, it's calm and clear. And no matter what the surface conditions are like, whether it's smooth or choppy or even a full-strength gale up there, deep in the ocean it's tranquil and serene. From the depth of the ocean you can look towards the surface and simply notice the activity there, just as from the depth of the mind you can look upward towards the waves, the brain waves at the surface of your mind, all that activity of mind—the thoughts, feelings, sensations, and memories. Enjoy this opportunity to just observe those activities at the surface of your mind.

At times it may be helpful to let your attention go back to the breath, and follow the breath to reground yourself in the tranquil place at the deepest depth of the mind. From this place it's possible to become aware of the activities of the mind without being swept away by them, to discern that those are not the totality of who you are; that you are more than just your thoughts, more than your feelings. You can have those thoughts and feelings and also be able to just notice them with the wisdom that they are not your identity. They are simply one part of your mind's experience. For some, naming the type of mental activity, like "feeling" or "thinking," "remembering" or "worrying," can help allow these activities of the mind to be noted as events that come and go. Let them gently float away and out of awareness. (Pause.)

I'll share one more image with you during this inward time. Perhaps you'll find it helpful and want to use it as well. Picture your mind as a wheel of awareness. Imagine a bicycle wheel where there is an outer rim and spokes that connect that rim to an inner hub. In this mind's wheel of awareness, anything that can come into our awareness is one of the infinite points on the rim. One sector of the rim might include what we become aware of through our five senses of touch, taste, smell, hearing, and sight, those senses that bring the outside world into our mind. Another sector of the rim is our inward sense of the body, the sensations in our limbs and our facial muscles, the feelings in the organs of our torso: our lungs, our heart, our intestines. All of the body brings its wisdom up into our mind, and this bodily sense, this sixth sense, if you will, is another of the elements to which we can bring our awareness. Other points on the rim are what the mind creates directly, such as thoughts and feelings, memories and perceptions, hopes and dreams. This segment of the rim of our mind is also available to our awareness. And this capacity to see the mind itself—our own mind as well as the minds of others—is what we might call our seventh sense. As we come to sense our connections with others, we perceive our relationships with the larger world, which perhaps constitutes yet another capacity, an eighth relational sense.

Now notice that we have a choice about where we send our attention. We can choose which point on the rim to visit. We may choose to pay attention to one of the five senses, or perhaps the feeling in our belly, and send a spoke there. Or we may choose to pay attention to a memory, and send a spoke to that area of the rim where input from our seventh sense is located. All of these spokes emanate from the depth of our mind, which is the hub of the wheel of awareness. And as we focus on the breath, we will find that the hub grows more spacious. As the hub expands, we develop the capacity to be receptive to whatever arises from the rim. We can give ourselves over to the spaciousness, to the luminous quality of the hub. It can receive any aspect of our experience, just as it is. Without preconceived ideas or judgments, this mindful awareness, this receptive attention, brings us into a tranquil place where we can be aware of and know all elements of our experience.

Like the calm depths of the sea inside, the hub of our wheel of awareness is a place of tranquillity, of safety, of openness and curiosity. It is from this safe and open place that we can explore the nature of the mind with equanimity, energy, and concentration. This hub of our mind is always available to us, right now. And it's from this hub that we enter a compassionate state of connection to ourselves, and feel compassion for others.

Let's focus on our breath for a few more moments, together, opening the spacious hub of our minds to the beauty and wonder of what is. (Pause.)

When you are ready you can take a more voluntary and perhaps deeper breath if you wish and get ready to gently let your eyes open, and we'll continue our dialogue together.

How was that? Some people have a tough time diving in; others feel at ease with the experience. If the breath doesn't work for you after a few sessions, you may want to find another form of mindful focus. Yoga or tai chi or walking meditation might be a more comfortable place for you to begin.

Just a few minutes a day of this or another basic mindful-awareness practice can make a big difference in people's lives. A number of my patients have reported feeling less anxiety, a deeper sense of clarity, safety, and security, and an improved sense of well-being. I hoped Jonathon would respond the same way.

Fortunately, Jonathon took to this exercise well and became committed to doing a mindfulness-of-the-breath meditation daily, initially for about five or ten minutes at a time. When his mind wandered from an awareness of his breath, he'd simply note this distraction and gently return his attention to his breathing.

The renowned psychologist William James once said, "The faculty of voluntarily bringing back a wandering attention, over and over again, is the very root of judgment, character, and will. . . . An education which should improve this faculty would be the education par excellence." Though James also said, "It is easier to define this ideal than to give practical directions for bringing it about," we actually do know how to refocus a wandering attention again and again—to use mindfulness practice to educate the mind itself. I truly did feel like a teacher for Jonathon, offering him an education in his own mind developed from 2,500 years of contemplative practice.

AWARENESS TRAINING AND STABILIZING THE MIND

As a part of his school's film club, Jonathon had been creating short documentaries exploring various parts of town with his parents' camcorder. He brought one of these projects in to show me early in our work together, and I was impressed by the creative ways he used camera angles to capture the mood and textures of this city in which both of us had been born and raised. His eyes sparkled with pride when he saw how much I enjoyed his creation. I told Jonathon about the metaphor of a camera on a tripod that I introduced in chapter 2. The lens of this camera is our ability to perceive the mind. Without a tripod to keep the lens steady, the mind can jump around like an amateur movie made with a handheld camera. Jonathon got it immediately— the blurry, bumpy film was like the feeling of being lost in his mood swings. Jonathon also liked the image of the ocean in the meditation exercise. He could identify with being a cork bobbing up and down on the surface of an agitated sea. But whichever metaphor of the mind works for you—wheel-and-hub, camera, sea—the sense is the same. There is a place deep within us that is observant, objective, and open. This is the receptive hub of the mind, the tranquil depth of the mental sea. From this place Jonathon could use the power of reflective awareness to alter the way his brain functioned and ultimately to change the structure of his brain.

Let's look at this process using the three legs of the mindsight tripod: observation, objectivity, and openness.

OBSERVATION

Jonathon first needed simply to become aware of his awareness, to observe how he focused his attention. As he discovered when he tried to focus on his breath, he would get distracted repeatedly and become lost in his thoughts, feelings, and memories. This is not doing the meditation "wrong." The point of the exercise is to notice these distractions—and then to refocus on the target (the breath), over and over again. Exercising attention is like developing a muscle: We bend our arm and then straighten it—flexing and relaxing our biceps, focusing and refocusing our attention when it wanders. This practice would not only develop Jonathon's ability to be aware of his awareness, but it would strengthen his attention to his intention—in this case, to focus on the breath. This monitoring of awareness and intention is at the heart of all mindfulness practices, from yoga to insight meditation, whether the focus is on posture and movement, the breath, a candle flame, or any of the myriad other targets found in the world's cultures. Bit by bit, Jonathon would build this mindfulness skill of "aim and sustain" and stabilize his mindsight lens.

In addition to his mindfulness exercise, Jonathon agreed to keep a journal of his daily activities, noting his shifts in mood, his mindful practice (or not), and his aerobic exercise. This was another opportunity to develop his capacity to observe his internal and external experiences and to reflect on the workings of his mind.

Recording his experience with mindfulness quickly revealed his lack of confidence in his mind. Nearly everyone who tries meditation discovers that thoughts and feelings keep interrupting our attempts at focus, even after years of practice. But intense feelings of frustration would flood Jonathon at such times, and he would write in his journal about how out of control this made him feel. He shared some entries with me where his self-disparagement bordered on not wanting to go on living. But there were glimmers of something else in the journal, as well: "My father told me to stop playing my music so loud and I blew up. He's so mean and doesn't know how to get off my back. . . . But tonight I could see my explosion at him like from a watchtower, sitting watching it fume, and it felt bad and I couldn't stop it." The next day, he said, he had calmed down, but he still felt that his mind had "betrayed" him again. "Only this time, I could see it instead of just being lost in it."

The observational distance that allows us to watch our own mental activity is an important first step towards regulating and stabilizing the mind. Jonathon was beginning to learn that he could "sit" in his prefrontal cortex and not get swept up by the brain waves crashing in on him from other neural regions. It was an important place to start.

OBJECTIVITY

If you're new to awareness training or meditation, you may find it helpful to compare it to what happens when you learn to play a musical instrument. Initially you focus on the characteristics of the instrument—the strings, the keys, the mouthpiece. Then you practice basic skills such as playing scales or strumming chords, focusing on one note at a time. This intentional and repeated practice is building a new capacity—it actually strengthens the parts of the brain that are required for this new behavior.

Awareness training is a skill-building practice in which the musical instrument is your mind. The aim-and-sustain skill developed during observation enables you to hold your attention steady, to stabilize the mind. The next step is to distinguish the quality of awareness from the object of attention.

We began this phase of Jonathon's awareness training with what is called a "body scan." During this practice, Jonathon would lie down on the floor and focus his awareness on whatever body part I mentioned. We would systematically move from his toes to his nose, pausing for him to take in the sensations of each region. When his attention strayed, his job was simply to gently note the distraction, let it go, and refocus—just as he'd done with the breath. What this immersion in body sensation was doing was directing his attention to a new area on the rim of his wheel of awareness. Sitting at the hub of the wheel, he could focus on the various sensations from his body, locating areas of tension or relaxation, and noting mental distractions while moving at will within this sixth-sense sector of the rim.

Next I taught Jonathon a walking meditation: twenty slow paces across the room with the focus of attention on the soles of his feet or lower legs. Same approach: When he noticed that his mind had pulled his attention away from the target, he simply refocused. These practices continued to build the aim-and-sustain function of observation, but they also were an entrée into objectivity: The focal point of attention changed with each practice, but the sensation of awareness remained the same. Awareness itself was becoming an expanded presence in his internal world.

Here is an entry Jonathon shared with me from his journal around this time: "Amazing realization: I can feel this change—my thoughts and feelings come up, sometimes big, sometimes bad—but they used to feel like who I was and now they're becoming more like an experience I'm having, not who I am, they don't define who I am." Another entry described an incident when he was upset with his brother. "I just got really mad . . . but then took myself outside for a walk. I was in the yard, and in the back of my head I could almost feel this split, something like a part of me that could see, and a part that could get lost in the feeling. It was really weird. I watched my breath, but I'm not sure that did much. Sometime later, I just seemed to calm down. It was as if I didn't take my own feelings so seriously."

During his home practice Jonathon was alternating among breath-awareness, body-scan, and walking meditation. But now his initial sense of frustration returned in a new form. He reported one day that he would get a huge "headache," a kind of "voice" that kept telling him what he ought to feel, what he should be doing, that he was doing his meditation all wrong, that he was no good.

All of these judgments were activities of his mind, I said, and I reassured him that he was certainly not alone—many of us have a judging voice that critiques our progress. But the next step in his growth would require him to stop being a slave to that voice. I felt this was a challenge that Jonathon was now ready to confront.

OPENNESS

Observation had enabled Jonathon to focus on the nature of intention and attention, the driving forces of mental life. Objectivity permitted him to distinguish awareness from mental activity, to further free his identity from the storms of his mental sea. But now that stormy rim activity was creeping back into his hub, in the form of the "shoulds" of expectations. These are the prisons of life. Trying to change how we actually feel by ordering ourselves to do so is a strategy that goes nowhere, fast. Open awareness is about accepting what is and not being swept up by those judging activities.

Does this seem ironic? Jonathon comes to me to try to change, and now I am encouraging him to accept himself as he is. But here is the distinction: Our effort to combat our actual experience creates internal tension, a kind of self-inflicted distress. But rather than march into our inner world and say "No—don't do that!" we can embrace what is and notice what happens. Amazingly, time after time people discover that letting things be also allows them to change. We can approach our inner world with openness and acceptance rather than with judgments and preconceptions. Consider this: If a friend came to you with some difficulties, you'd probably listen to her first, invite her to bring up whatever came to mind, and offer her an open heart and a shoulder she could lean on. This is what openness entails—attuning to what is, being kind and supportive to ourselves, letting our state be receptive rather than reactive.

Jonathon, however, had not yet learned to be kind to himself. He'd be focusing on his breath, for example, and if he got distracted by some memory of last weekend, some concern about schoolwork, or thoughts about a fight with a friend, then he'd get a "sense" in his head that he was "not meditating right" and that he was "not a good meditator." I suggested to Jonathon that these harsh self-criticisms were just another mental activity for him to notice. They were judging thoughts, I told him, and when they came up he could try simply labeling them—"judging . . . judging . . ."—and then bring his attention away from them and back to his breath. Jonathon decided he preferred using the label "doubting, doubting" to remind himself of the undermining nature of these distracting thoughts.

The quality of openness is the third tripod leg stabilizing our mindsight lens. It means that instead of being swept up by shoulds, we come to accept ourselves and our experiences. But to get to this place of inner attunement, of internal acceptance, we must first become aware of when we are our own prison wardens.

A STABILIZED MIND

Jonathon noticed the changes that were emerging. He would go for a run or ride a bike during stormy times, trying to find some way out of the mood that seemed to take him over. These rhythmic physical activities helped him to calm his body, to get grounded in his awareness, and to bring himself back into balance. As the weeks unfolded, Jonathon described a new experience. He began to sense his raging thoughts and intense emotional storms with more clarity, seeing them but somehow not becoming swept up by them. What surprised him, and thrilled his parents, was that he seemed to find a new way to actually calm the storms.

This is what Jonathon wrote in a journal entry one night: "I had a fight with my Mum this afternoon and I went to my room before dinner. I thought of killing myself. There it is again. This will never get better. Just when I think things are changing, they stay the same. I was late coming home from school and she just laid into me, she was SO angry. . . . I sat in my bed and just thought—what's the point. But then the feeling of being absolutely helpless seemed to float in my head, like a raft or a boat, some kind of log or something. But instead of the usual feeling of being on that boat, floating away, I was somewhere else. I could see that the raft was just a feeling, just the feeling of me not being able to DO anything to get out of this. And what was really weird was that once I let the boat just be there, kind of in my head but separate from 'me,' not being on it, it didn't make me feel so bad. Then when I looked at it straight on, like just some kind of helplessness, it just disappeared." In the session that day, Jonathon and I spoke about how this experience of the "boat" let him see that in fact he did not have to just float aimlessly on that feeling of despair. He had learned that he could do something to prevent being ambushed by his feelings. Jonathon also learned that just observing his own inner world with acceptance had a strong soothing effect on his distress. He told me that he began to notice he could soften the violence of his thoughts and feelings by looking directly at them and not running from them. Understanding that he could actually reverse the flow of his feelings and thoughts gave him wonderfully positive feedback about his own abilities. In many ways, Jonathon's experiences echoed the research finding that people with mindful awareness training have a shift in their brains towards an "approach state" that allows them to move towards rather than away from challenging situations. This is the brain signature of resilience.

Later on Jonathon wrote, "I know this sounds lame, but my view of life is changed now. What before I thought was my identity I now realize is just an experience. And being filled with big feelings is just some way my brain gives me experiences but they don't have to say who I am."

I was moved by his discoveries, and in awe of his ability to articulate such deep insights. Now we had to see how he could refine this newly enriched monitoring ability to begin to alter the way energy and information were flowing in his internal world—to stop his mind from being flooded with those "big feelings" in the first place. Having already learned how to use self-observational skills to see his internal storms, he was now ready to learn techniques that would enable him to do something about them. I next taught Jonathon basic relaxation skills, inviting him to imagine a peaceful place from his memory or imagination that he could evoke at times of distress. We practiced this imagery in the safety of the office and combined it with the grounding feeling he'd get by just noticing his body in the chair or sensing his breath. These relaxation and internal imagery techniques would provide him with some readily accessible ways of calming himself. Over time, Jonathon learned to ward off an impending "low-road" meltdown by noticing his change in bodily state—his pounding heart, churning belly, tense fists—and then the very act of noticing would soothe him. Jonathon was experiencing the power of a stabilized awareness of the mind to achieve mental equilibrium.

In our sessions as the months unfolded, Jonathon became more and more confident of his ability to look inward and then to change what was going on. In his journal he wrote, "I am beginning to see how my own way of paying attention to my feelings changes what they do to me. They used to explode and last for hours. Now after a few minutes, I can see how they can crash around and then, as I don't take them so personally, they just melt away. It's strange but I'm starting to believe in myself, maybe for the first time."

Change required the ability to accept what was there and have the strength to let it be, until his mind became stable again. He and I both knew how hard this road had been for him. The storms of his life had been a huge challenge, but they also provided the motivation for him to find a way to create a harbor of safety in his own mind.

What had changed for Jonathon? We don't have the brain scans to say for sure from a neural point of view—but what I picture is that over these hardworking months of twice-weekly sessions and essentially daily awareness practice and aerobic exercise, Jonathon was growing his middle prefrontal integrative fibers. His new way of focusing his attention, of integrating his consciousness, would have been made possible by his middle prefrontal areas expanding their connections and beginning to grow the GABA inhibitory fibers that could calm his subcortical storms. The "GABA-goo" could then soothe his irritable limbic amygdala so that it didn't recruit his brain-stem areas into the fight-flight-freeze routine that had been driving Jonathon mad. He was also likely moving more towards a "left-shift" brain state of approach. With this new integration Jonathon was learning how to coordinate and balance the firing of his brain in new and more adaptive ways. He could now "sit" in the sanctuary of his newfound awareness without being swept up by the mental activities that used to overwhelm him. This mental training was more than just a way to alleviate his roller-coaster symptoms—it was a way for Jonathon to become more resilient, and more himself. "I'm feeling almost like a different person—like I'm stronger now. I don't want to say this too much to jinx it, but I feel really good—really clear."

By six months into our work together, most of Jonathon's symptoms of emotional turmoil appeared to have dissipated. Sitting with him in the room had a different feeling: He seemed more at ease, clear, and lighthearted. He seemed more comfortable in his own skin.

"I just don't take all those feelings and thoughts so seriously—and they don't take me on such a wild ride anymore!" We continued to work on his practice and solidified his newfound skills. On our last visit, after a year of therapy, Jonathon stood up to shake my hand and I saw again that sparkle in his eyes that had so often been hidden behind a mask of anguish and fear. Now his gaze was clear, his face at ease, and his handshake confident and strong. He must have grown at least three inches since he first stepped into my office, what felt like ages ago.

After high school, Jonathon moved on to attend college out of town. It's now years later, and I recently ran into Jonathon's parents at a neighborhood store. They told me that he is "doing great" and has not had a recurrence of his roller-coaster mind. He's studying film, and psychology.

6

Half a Brain in Hiding
Balancing Left and Right

STUART HAD JUST CELEBRATED his ninety-second birthday when his son brought him to see me. "I've never needed a shrink, and I certainly don't need one now," he announced as he refused his son's arm and walked with deliberate steps into my office. Stuart didn't look a day over seventy. He was a handsome man, clean-shaven, with a head full of gray, wavy hair combed neatly just over his ears. "I'm here because of my son," he added. "It's a stupid idea, but he thinks I need help."

Randy had told me on the phone that his father was depressed. He'd read a newspaper article about depression in the elderly, and he'd concluded that Stuart's depression had been triggered when Randy's mother, Adrienne, was hospitalized for pneumonia six months earlier. Stuart and Adrienne had been married for sixty-two years, and after she returned home, Stuart had become, in Randy's words, "a basket case." He'd stopped going into his old law firm several times a week. He'd stopped going for walks and seeing friends. He'd stopped calling Randy or his brother on the phone. And although he'd never been very involved with his grandchildren, he was now even more removed. At family events, he'd sit off to one side reading a newspaper or watching the news. Even at home with Adrienne, Stuart appeared uninterested and withdrawn.

But as Stuart and I began to talk after Randy stepped out, what first struck me was not so much depression as a kind of emptiness. Stuart did seem flat; his tone of voice lacked variation and his face lacked expression. He narrated the details of the last six months as if they were from some television show that happened to be on when he was waiting for the news. He was energized and alert but seemed distant and dispassionate.

I looked into Stuart's eyes, searching my own feelings and sensations for some mirrored sense of what might be going on in him. As I've discussed, we use our whole body as our mindsight "eyes," and I was conscious primarily of a dulled, bland sense of something missing. You've probably noticed that when you are with someone who is depressed, you begin to feel down yourself—heavy, sad, distant, and alone. But with Stuart at first I felt nothing, and then I picked up a vague sense of fear, a hidden apprehension. Was this my own fear of meeting someone in his nineties whom I might not be able to help—who, in fact, had just declared I could not? Was I simply projecting my own fear of getting older, of illness, of loss? Or was this in fact my resonance circuitry accurately reflecting something going on in Stuart?

After a few minutes, he seemed to settle into his chair and feel more comfortable just being there as we "chatted." I found out more about Stuart's life—his work as an intellectual property lawyer, his favorite football and cricket teams, his educational history, and how he met Adrienne. He had retired just ten years earlier from his partnership at a local law firm, and he told me that he'd continued to consult on cases there, enjoying his status as the wise elder. He'd gone in to meetings even while Adrienne was in the hospital. But now, he acknowledged, he was staying home and reading a lot. Other than that, "things were fine." As he spoke, I watched for signs of early dementia and found none. Stuart's memory, his attention, his orientation to reality, all seemed intact.

Then I asked him how he'd felt when Adrienne was ill. "I know this does not sound correct," he replied, "but to tell you the truth, I didn't particularly worry. She had the best doctors, and they said she was going to be fine. You know," he went on, "even when one of my law partners was diagnosed with lymphoma, I felt nothing. People get sick, they die. That's it. I know I should feel something about it, but I just don't."

Stuart's saying that it didn't sound "correct" caught my attention. Somehow he realized that his reactions were not quite normal, and he seemed to be reaching for a category—"correct" or "incorrect"—to understand them. Maybe I could align myself with that awareness, that curiosity about other possibilities for feeling. I wondered what had led him to be so stuck and disconnected and what we could do about it.

Near the end of the session, I asked Randy to join us again in my office. Both Stuart and his son agreed that he had always had an "even-keel" temperament. They acknowledged his "feistiness," and the "no-nonsense" way he told people what he thought, but neither of them recalled a time when Stuart had truly lost his temper. Nor were there any extended periods when he had been sullen and moody or, in contrast, elated. All in all, as Randy put it, Stuart had been "the Rock of Gibraltar in everyone's life." Though Stuart didn't respond to this, the glimmer in his eyes gave me the sense that he cared deeply about his son. It also gave me hope that I could help him, and I was relieved when he agreed to return for "a few" sessions.

FROM PAST TO PRESENT

Stuart did come back, as prickly as ever. When I started to ask him for recollections of his childhood, he told me I was being "ridiculous." Didn't I realize, he said, that at ninety-two his childhood was certainly a "moot point"? "Why go into that now? I knew you therapists were out of your minds."

I felt like saying "objection overruled," but I restrained myself. Humor can be an important way to connect, and it may even stimulate neuroplasticity, but it didn't seem right at this point. Instead I told him that to get into his mind, it would in scientific fact be useful to review his recollections of his past. You may imagine Stuart-the-attorney's response: "I don't need help. So that is irrelevant."

I use my interview questions for two purposes: One is to obtain details of a person's life events. The second is to get a sense of how the story is told. I was looking for developmental challenges to which he needed to adapt, such as losses or trauma. Our personality emerges as our inborn, often genetically influenced temperament—such as shyness or moodiness—interacts with our parents, peers, and teachers and with our experiences at home and at school. Random events—in the womb, in our early years, even later—also contribute in unpredictable ways to how we develop. We adapt to all that we are given and to all that we encounter. We know nothing else. We do the best we can, and our sense of self emerges in a dance among innate characteristics, adaptations to experience, and just plain chance.

Once I got Stuart talking, his memory was excellent for details about the town where he grew up, the games he played as a child, the make and model of his first car, and even the historic and political events of the time. But when it came to my questions about his early family life—or any family life—his responses were consistently vague. "My mother was normal. She ran the home. My father worked. I think my brothers and I were fine." To a question about how his family life affected his development, Stuart responded, "It didn't. . . . My parents gave me a good education. What's the next question?"

Stuart insisted that his childhood was "fine" even though he said that he did not remember the details of his relationships with his parents or two brothers. He insisted that he "just didn't recall" what they had done at home, what life *felt like* for him as a youngster. The details he gave me sounded like facts, not like lived experience. This was true even when he told me that he had been with his brother during a bad skiing accident, which had resulted in the loss of his brother's leg. His brother had recovered and was "fine."

This challenging session gave me some important information. Stuart's generalized recollections, his lack of recall for family experiences, his insistence that these relationships did not impact his life, are all classic findings of a certain kind of autobiographical narrative that I had been studying for years. A vast amount of research suggests that such a narrative develops from being raised in a home where emotional warmth is absent.

This was confirmed when Adrienne came in with Stuart the following week. She said Stuart's parents were "the coldest people" she'd "ever met on the planet. You just can't imagine why people would be so odd, so frozen, so uncaring . . . poor Stu." At eighty-three, Adrienne herself was in great form, and she looked at Stuart with pride and affection. When she turned to me, she said that she hoped I could help him "break out of his shell."

Adrienne's comments reinforced my impression that while Stuart had become even more removed since her pneumonia, he had always been emotionally distant. But something had happened to him when Adrienne was hospitalized, something that had hit him in a way he couldn't, or wouldn't, discuss. He seemed to Adrienne to have lost interest in their life together, actively withdrawing to his world of history books and law journals. She said that she hoped the therapy could make him "happier." Stuart remarked that he didn't know exactly what that meant, but he felt that his wife deserved to have a better companion in their retirement. He agreed to continue therapy for three or four months to see what he and I could do together.

RIGHT AND LEFT

Perhaps it was being raised by "the coldest people on earth" that had resulted in Stuart's underlying rigidity, perhaps it was the genes he inherited, or perhaps there was some still-unknown factor. We didn't have to know for sure to intervene. That's the beauty of the integrative approach. We can move the system towards a FACES flow by focusing on the three points of the triangle of well-being: mind, brain, and relationships. The basic questions: What is going on now? What can be done to promote differentiation and linkage?

To understand Stuart, let's explore how the left and right sides of the brain present us with quite different ways of perceiving reality and of communicating with one another. These differences form the neural underpinnings for the kind of responses Stuart had given during our developmental interview. In those with a cold, emotionally empty childhood, one side of the brain often seems to be understimulated while the other becomes excessively dominant. Stuart's style of narrating facts without autobiographical detail, and his lifelong engagement in a professional life that required highly logical intellectual work but little in the way of connecting with people emotionally, strongly suggested that he had a dominant left brain and an underdeveloped right brain. The left and right brain are in the news these days, and I even hear them being discussed at social gatherings (at least at the parties I attend). But it's easy to oversimplify the differences, so let's take a closer look.

From the beginning of life we communicate with one another in the "nonverbal" realm. We send and receive signals through our facial expressions, tone of voice, posture, gestures, and the timing and intensity of response. When we were babies, nonverbal signals were our lifeline, the only way we could convey our needs and wants. We cried, we flailed our arms and legs, we frowned or turned away when we were hungry, scared, in pain, or lonely. We smiled, cooed, and snuggled into our comforting caregiver if we felt safe, secure, and well fed. And as our caregiver responded to our signals we were linked together by these nonverbal patterns of energy and information. This is how Leanne "felt felt" by Barbara before her accident. This is how many of us became a "we" with our parents.

These nonverbal signals are both created and perceived by the right side of the brain, and neuroscientists have found that the right hemisphere is the more developed and more active during the first years of life. It was the scarcity of such nonverbal signals that I noticed in my first few sessions with Stuart. Here was a bright, articulate, accomplished professional, but he seemed to lack these fundamental textures of conversational life. Of course we also connect with one another through words, such as those you are reading at this moment or those Stuart deployed so skillfully in his career. Words become much more important after the early years—and this is when the left hemisphere becomes more active. Throughout our childhood and adolescence, the right and left hemispheres go through cyclical phases of growth and development.

The right hemisphere is more directly connected to the subcortical areas of the brain. Information flows from body to brainstem to limbic areas to right cortex. The left hemisphere is more removed from these raw subcortical sources—from our physical sensations, our brainstem survival reactions, and our limbic feelings and attachments.

Because of this developmental history and anatomical configuration, our right hemisphere gives us a more direct sense of the whole body, our waves and tides of emotion, and the pictures of lived experience that make up our autobiographical memory. The right brain is the seat of our emotional and social selves. We create images of our own mind and that of others using right cortical real estate; the right side also has a greater role in coping with stress and regulating the activity of the subcortical regions. But these are not absolute distinctions. Normal life weaves these right-dominant activities into the equally important, but different, left-dominant information flow.

Even to communicate these ideas, I need to use my more conceptual, fact-based, analytical left side—and you need to use yours to understand them. The left hemisphere, being less directly influenced by the subcortical happenings below it, lives in a kind of "ivory tower" of ideas and rational thought compared with its more visceral and emotional right-hemisphere counterpart. But the two spheres do communicate. These right-left cortical neighbors are linked by the corpus callosum, a band of neurons deep in the brain that enables energy and information to be sent back and forth between them. Considered in isolation, these differing patterns of energy and information flow enable us to have something like "two minds" that can cooperate or compete. We'll call these the right and left "modes." When the two hemispheres collaborate, we achieve "bilateral" or "horizontal" integration.

Your left hemisphere loves linear, linguistic, logical, and literal communication. Also a list maker, the left loves to label things. It specializes in syllogistic reasoning, using chains of logic, and identifying cause-effect relationships. We know that the left hemisphere is coming online when two- and three-year-old children start asking "Why? Why? Why?"

Here's a thumbnail sketch of each mode: Left—Later developing, Linear, Linguistic, Logical, Literal, Labels, and Lists. Right—Early developing, Holistic, Nonverbal, Images, Metaphors, Whole Body Sense, Raw Emotion, Stress Reduction, and Autobiographical Memory.

Another way of thinking about the two modes is that the left is more "digital," with an on-off, up-down, right-wrong categorization of information, while the right is more "analogic." Brain anatomy reveals a possible reason for these differences in the contrasting micro-architecture of the two regions.

The right mode creates an "AND" stance, while the left creates an "OR" point of view. Using my right mode, I see a world full of interconnecting possibilities: This AND that can be true. And together, wow, they could make something new! Using my left mode, I see a world more divided: Is this OR that true? For the left, only one view can accurately reflect reality. And when I'm looking at the world through my left-mode OR lens, I have no sense that I'm choosing to see the world this way. It *is* the way. And the other way, the right mode, well, it is just plain *wrong*.

I've worked with couples where one person is dominant in the left mode and the other dominant in the right mode. Person A says that he feels sad. Person B responds, "You're nuts. There's nothing to be sad about." A looks perplexed but cannot speak. B seems to feel she's won. What kind of game is that . . . sad, disconnected, certainly not integrated . . . a lose-lose situation.

You might be wondering if, having two powerful but quite distinct neural processors in our heads, we must inevitably be at war within ourselves—as well as with people who have a mode dominance that is different from ours, as in the couple described above? Sometimes we are. When one mode dominates the other for long periods, rigidity and/or chaos result. This is how Stuart felt to me in the office—rigid and disconnected.

There are many reasons that someone might grow up "leaning to the left." What if our need to be close to others—to share our nonverbal signals, to feel seen and safe—is not met by a caring, connecting, communicating other? Or even worse, what if those early interactions are terrifying? How can we live with that sense of uncertainty? If we are living in an emotional desert or are being tossed about by violent storms, our right hemisphere may shrivel in response. Retreating to a more left-dominant mode puts our awareness in a safer place. It is one common and adaptive strategy to survive. But there are better ways, and I was hoping I could help Stuart discover them.

SNAGING THE BRAIN

Was starting therapy with a ninety-two-year-old like "trying to teach an old dog new tricks"? If Stuart's right-mode neural circuits had been dormant for decades—for nearly a century—could we stimulate them to become active? Whether this dormancy was from experience, genetics, chance, or all three, was there a way to change his present neural functioning? And if we activated these circuits, could we actually expect them to make new synaptic connections, or to even grow new integrative neurons? The science of neuroplasticity, together with clinical work in neural rehabilitation, suggested that yes, it might be possible. And this is what I told Stuart.

I drew a picture of the brain and told Stuart about the left and right sides. Our goal, I said, was to help him develop a more balanced whole brain—to add new right-hemisphere abilities and reinforce his already well-developed left side. Then I introduced one of my acronyms, SNAG, for Stimulate Neuronal Activation and Growth. I said we could SNAG his brain to create and strengthen neural connections. Wherever neural firing occurs, existing neurons can make new or enhanced synaptic connections through the process called *synaptogenesis*. New neurons can be stimulated to develop, as well—a process called *neurogenesis*. I also explained how the myelin can thicken, which increases the electrical conduction among interconnected neurons. And, as I'd told Jonathon, among the keys to neuronal growth are novelty, attention, and aerobic exercise. Stuart liked the SNAG acronym, and I was happy his left mode seemed to appreciate the wordplay.

Short of using an electrical probe, how can we strategically target a particular portion of the brain? The answer is attention. When we focus repeatedly on specific skills, moment-to-moment neural activity can gradually become an established trait through the power of neuroplasticity. First we would use the focus of Stuart's attention to SNAG his right hemisphere, and we would do this by working on skills that would help enhance differentiation by stimulating growth of that region. The work on linkage of the right to the left would follow.

I wanted to be sure Stuart's left mode was game for this process. I talked about what we were doing as skill building, and I told him about research on how the brain changes in learning to play a musical instrument. If the instrument is a violin, as I mentioned earlier, studies have demonstrated that attention to those intricate left-hand fingerings will build the part of the cortex that regulates the sensations and motor control of the left hand. Those areas will be much larger than areas regulating the right hand, which is using the bow in less subtle ways.

I then told Stuart that he and I could focus his attention in specific ways to allow his right hemisphere to become better developed. We simply needed to give ourselves the biological time necessary to enable these new synapses and neurons to grow and to become part of a newly integrated system.

DEVELOPING THE RIGHT

Stuart was attracted to the logical and scientific discussions of the brain that I shared with him as part of our treatment plan. Nothing I said put him in the wrong. I emphasized how the brain responds to experience early in life, how we adapt and "do the best we can" throughout the lifespan. Whatever the factors involved, he could now change, if he wanted to, through experientially induced development. The final point I made was that we were not trying to change who he was, but to expand his potential by nurturing an underdeveloped set of circuits in his brain. I hoped that framing our work in this way would give Stuart enough of a feeling of safety about the ultimate goal of the work—which was to open himself up to emotion and allow himself to become vulnerable—that he would be willing to dive in.

As our session ended, Stuart paused for a moment and reminded me of how he'd "felt nothing" when Adrienne and his law partner had gotten sick. He then said something I'll never forget: "I know people say they feel this or feel that . . . but in my life, I basically feel nothing. I really don't know what people are talking about. I'd like to know before I die." I knew then that he and I could work together to rewire his brain. Both of us seemed up for the task.

BODY SENSATIONS

Since Stuart himself had admitted his feelings were inaccessible, we began with something more tangible: the body.

To tap into this somatic sense, I led him in the kind of body scan I'd done with Jonathon, starting with the right foot and moving up the leg. You may recall from "The Brain in the Palm of Your Hand" (page 14) that the right side of the body is represented in the left hemisphere, and vice versa. In fact, an image of the whole body is mapped only in the right, but I wanted to start on the side of the brain where Stuart would feel most competent. After he had successfully focused his attention on the right side, we did the same scan for the left leg. It turned out that he could locate sensations on either side without a problem. This suggested that the basic neural circuitry connecting one side of the body to the opposite side of the brain was intact.

But when I asked Stuart to place the sensations of both legs simultaneously "in the front of his mind" he faltered. "I can't really see it. It's like a shimmering object. First I see one side, then the other." So basic functioning was intact on both sides, but he could not integrate the two sides simultaneously within his awareness. We worked on this ability throughout the remainder of the full body scan, as I repeatedly asked him to focus his attention on first one, then the other, then both sides at once.

When I turned the scan inward and asked Stuart to sense his body's organs, he had an even harder time. Research suggests that interoception—our perception of our internal bodily state—is mediated primarily via the right brain. As I discussed in "Riding the Resonance Circuits" (page 59), we pass the signals of our body's interior—and also of our limbic states—through the conduit called the insula up into the right middle prefrontal regions of the brain. I couldn't tell whether Stuart's difficulty with interoception was due to lack of development of the resonance circuitry or to a lack of linkage with his left-side linguistic centers, which would be needed to translate such interoception into words. Whichever the cause, like any skill, focusing his mind on this sensation would become easier with repeated practice, and I didn't want to let him get too frustrated. It was time to move on to another aspect of the right mode.

NONVERBAL CONNECTION

We first develop relationships during our early years, when our right hemisphere is dominant in both growth and activity. Perhaps this is one reason the right mode specializes in close relationships through-out our lives. The right side also specializes in self-soothing. Babies use their left hand (controlled by the right brain) for comforting, whereas the right tends to move out to explore the world. Brain imaging has revealed as well that left prefrontal activation appears to be associated with an approach state, in which we seek out and open ourselves to new experiences. Right frontal activation, on the other hand, is associated with withdrawal, a turning inward away from novelty. Interestingly, social display rules—the codes that tell us how we are supposed to act in groups—are a left-mode specialty. So over-all, the left seems more outwardly focused, while the right is an inte-rior specialist, exploring our own and others' internal worlds. This could help explain why Stuart thrived in settings such as the court and conference room but faltered when it came to more intimate relationships.

To introduce Stuart to the rich inner world of reflection and rela-tionship, I engaged him in a series of nonverbal communication "games." At first they were simple: I'd make a facial expression and he'd name the emotion—sadness, fear, anger. Then I asked him to imitate my expression. He refused to try until I told him the reason for the exercise, and even then he could not do it. But after a few ses-sions he got pretty good at imitating me. For homework, I had him watch television shows with the sound off. This would engage his right hemisphere's nonverbal perception ability—and perhaps bore his left hemisphere into relaxing, if not falling asleep outright. (When we were together, I had to be careful not to engage him too often with my own left hemisphere. He loved explanations, asked me many questions about the research, and enticed me with fasci-nating stories about other topics. But we had work to do, and we had to meet, right brain to right brain.) When we engaged in nonverbal "games" together, it felt like this play was the brain food that Stuart had been waiting a lifetime to receive.

IMAGERY

As our nonverbal communication improved and Stuart became more attuned to his own body's sensations, I decided it was time to explore his internal world of images and autobiographical reflection. I asked him to recall the evening before our session and his breakfast that morning, and to convey his recollections as images rather than facts. On the one hand, this was safe territory for Stuart. These would be neutral memories, pictures of his own experiences in the recent past. But here was the tricky part: Autobiographical representations are right-mode dominant, they are not in word form. So I would say, "Just notice what emerges in your awareness." "Just notice" invites a broad sensory experience that is closer to the right-mode flow of images before words. Stuart wanted to summarize and evaluate: "I had a good evening." "I had cornflakes for breakfast." What came hard to him was telling me "I scoop the cornflakes into my blue bowl and hear the dry sound they make. The milk carton feels cool in my hand, and I pour it slowly until I see the milk almost covering the flakes. I sit down and I notice that the sunlight is in my eyes."

Stuart and I went on to images of neutral scenes, such as his favorite beach, his yard at home, his last vacation. Again these images did not come easily into his awareness. Verbal concepts dominated his mental landscape, and he would start to explain—not describe—where he went and what he did on his vacation. But Stuart loved a challenge, and he slowly learned that the mind's activities are not just those linguistic packets of words we share in school and business—the mode he had been so rewarded for throughout his youth and adulthood.

Of course you've noticed the paradox—we were using words to access the wordless right-hemisphere realm of sensations, images, and feelings. Aren't words the left brain's specialty? Yes and no. When we *explain* a science experiment or a legal proceeding, we are relying heavily on the left. When we *describe* rather than explain, we are bringing the experientially rich right side into collaboration with the word-smithing left hemisphere. The challenge was to invite Stuart's left to participate but let the right stay strong. This would be the beginning of a more balanced linkage of left and right.

With reassurance and practice, what began as fleeting images became a more steady film in Stuart's mind's eye. He slowly became immersed in the sea inside. Over a period of weekly visits spanning several months, Stuart began to enjoy what at first had been a frustrating exercise. For homework, I gave him a book about learning to draw using the right side of the brain. He also began writing in a journal for the first time in his life. Sometimes he'd bring his entries in for us to read together—reflections on how he was changing, on the new world that was opening up to him. At times he'd write about how uncertain he felt, sometimes feeling afraid that he "couldn't do this" and that he was no good at describing, a "failure at feeling." But as time went on, he said that he had "a whole new way of seeing." The key for him, he said, was adjusting to the reality that he could not control where his images would take him. How different this must have been from studying and practicing law. Once he could relax his left-hemisphere predilection for control and certainty, his mind could become free to open to his inner world.

MAKING THE LINK BETWEEN LEFT AND RIGHT

Finally Stuart and I moved to the level of feelings. Stuart's initial statement, "I don't know what I feel," had slowly given way to his being able to articulate how the muscles in his arms felt, where his face was tense, when his chest felt heavy, or the uneasiness in his belly. From such bodily sensations, he'd sometimes become aware of images—a picture in his head of being with someone, or of hiding or running away. Tuning in to the body's signals and to the imagery that arose from them also helped Stuart gain awareness of his feelings, because feelings themselves are the subjective sensation of what is going on inside the entire body—from our limbs and torso up to our brainstem, limbic areas, and cortex. However, it was still difficult for him to translate these sensations, images, and feelings into words when I asked him about them.

Stuart was not alone in this; finding the words to accurately depict our wordless internal world is a lifelong challenge for many of us. Poets offer us a window into the mastery of this neural skill, but few of us have a poet's gift for translating feeling into words—and it really is quite a feat of a translation, if you pause to think about it. We use our left hemisphere's linguistic packets to ask another person's left hemisphere a question about his experiences or feelings (or to ask ourselves the same question). That person must decode those signals and send a message across the corpus callosum to activate the right hemisphere, which comes up with the nonverbal somatic-sensory images that are the "stuff" of feelings. He then has to reverse the process, translating the right hemisphere's internal music back into the digital neural processors of the left hemisphere's language centers. Then a sentence is spoken. Amazing.

This was why it was important for Stuart to write in his journal and to make it a record not only of his thoughts but also of the sensations, imagery, and feelings that were entering his awareness. As our weekly practices continued, Stuart's journal revealed the increasingly intricate world of his right mode, replete with dream descriptions, poems, and heartfelt reflections on his life. He seemed to enjoy reflecting on the internal world that had now become accessible to him.

Using words to describe and label this internal world can actually be useful not just for those like Stuart who have trouble accessing their emotions, but for those who need to find a way to bring balance to overactive feelings. Such people have an excess of right-mode flow without enough linkage to the left (versus Stuart's excess of left-mode activity without enough linkage to the right) and may suffer from emotional dysregulation and chaotic outbursts. They can become overwhelmed by fragmented autobiographical images, filled with bodily sensations, awash in emotions that overwhelm and confuse. For these people, balance entails gaining some mental distance in the sanctuary of the left mode. Since the right hemisphere is more intimately linked to the emotion-generating subcortical areas, we can see why raw, spontaneous feeling is more fully and immediately felt in the right mode—and why it makes sense that linking the right and left modes through the left-hemisphere function of language might bring about the necessary balance. And indeed, studies done by my colleagues at UCLA have actually shown that naming an affect soothes limbic firing. Sometimes we need to "name it to tame it." We can use the left language centers to calm the excessively firing right emotional areas. But, again, the key is to link left and right, not replace one imbalance with another.

BUILDING THE "WE" OF MINDSIGHT

One day Stuart mentioned that his oldest grandson had broken his leg skiing. I recalled the early session when he'd told me about his older brother's skiing accident, and I wondered if he had some unresolved emotions about it—which he might be open to discussing at this point. When I brought up the topic, Stuart became tearful, and I thought I had hit a tender spot in his memories. I said that perhaps the event was still quite raw in his mind.

Stuart shook his head. "No, that's not it," he said, wiping a tear from his cheek.

"What do you think it is?" I asked him, puzzled about what could be creating this new and, for him, unusually intense emotional response.

"It's not about my brother, or about the accident," he said, looking straight at me. "It's that I can't believe you remembered what I said months ago. . . . I can't believe you really know me."

We sat in a powerful silence, looking at each other. I felt his presence in a way I had never experienced with him before. We talked about that sense of connection between us and about some other things on his mind, and the session ended. When he rose from the chair, he came over to me and shook my hand, then brought his left hand up to cover our clasped right hands. "Thanks," he said. "Thank you so much for everything. This was such a good session."

I can't really put words to what happened, but—half a year into therapy—there now seemed to be a "we" in the room. If we had had brain monitors on hand, I think they would have picked up the resonance between us. Just as Stuart had been moved to tears at realizing that his mind was in mine, I felt deeply moved by feeling, for the first time, that mine was in his. There was a deep and open connection between us.

STRENGTHENING SYNAPTIC INTEGRATION

A cascade of positive effects seems to emerge spontaneously when integration has been initiated. It's like the old physics idea of pushing a ball up a hill to get it rolling down the other side. It takes considerable effort and deliberate attention to move beyond the initial engrained, nonintegrated state—to push the ball up the hill. This is the intentional work of change. But ultimately the emerging mind takes its natural course towards integration, and the ball flows effortlessly down into the valley of coherence. Integration is the mind's natural state.

In the beginning of our work together I had thought I would need to cultivate empathy in Stuart step by step, starting with the basics of taking in others' emotional communication and then responding compassionately. Before he could make mindsight maps of others' minds, he'd need to learn how to open up to resonating with their emotional states and then how to access those states in himself. But I realized in retrospect that we had already worked on those basic building blocks. Our focus on his bodily sensations built interoception; reflection and journal writing opened his awareness to feelings; and imagery work strengthened his ability to attend to nonverbal experiences. These essential elements of empathy are all forms of integration. Once we got the ball rolling, Stuart's wonderful and now-eager mind was ready to do what it was born to do—to connect with others, and with himself.

Nine months after our first meeting, I received a call from Adrienne. She asked me if I had "given Stuart a brain transplant." She told me she was stunned at how tuned in to her feelings he had become, and that they were happier now than they'd ever been. She wanted to share what had happened the night before. She was standing next to Stuart as they said good night to a guest, and she put one hand on his shoulder. In the past he would have stiffened a bit or moved away, but instead he said, "Wow, that feels good." Then he actually let her give him a shoulder massage—for the first time in sixty-two years of marriage.

The next time I saw Stuart, he reflected on how important Adrienne was to him. He had come to understand that his parents' coldness must have been so painful that he just retreated into his schoolwork, then into his profession, and lost touch with others and with himself. When Adrienne was ill, he pulled away even further. Now he could become aware that the fear of losing someone who had loved him for so long had felt unbearable. We began to work in therapy on facing life's fragility, of learning how to care deeply yet come to accept that we cannot control how our lives, or our relationships, unfold. "I know it's easier to hide in books," Stuart had written in his journal, "but they just don't feel as good as love."

Without my mentioning it, Stuart brought up the moment when Adrienne touched his shoulder. "I think I just never wanted to feel that I needed her. It was just easier—for all these years—to not need anyone. How hard this must have been for her . . . and I am so grateful she stayed with me all this time. She said she liked massaging my shoulders, even if it took me two-thirds of a century to say 'yes'!" As for Stuart, the twinkle in his eyes said it all when he told me that the massage "felt great!"

A year after our last session, as he approached his ninety-fourth birthday, Stuart sent me a note: "I cannot tell you how much fun I am having. Life has new meaning now. Thank you." I thank him for teaching me, for teaching all of us, how resilient our integrative brains can be.

7

Cut Off from the Neck Down
Reconnecting the Mind and the Body

ANNE'S FIRST VISIT was on a rare rainy day in Los Angeles. She must not have brought an umbrella, because her long black hair was soaked. It was bundled into a loose knot at the side of her head, and a stream of moisture was quickly darkening the shoulder and neckline of her jacket. I couldn't help watching the dark spot spread, but Anne didn't seem to be bothered. I'd soon learn that this lack of interest in her body was more than just a passing state of being caught off guard in the rain.

Anne looked around the room, slouched back into the couch, and said, "Well, here I am, but I'm not sure why." Anne was a forty-seven-year-old doctor and the mother of eleven-year-old twin girls. She told me that she had been putting off going to her doctor for a follow-up exam for more than a year. Her slightly raised blood pressure and some findings during a routine heart exam had concerned him, and he'd asked her to return a few weeks later, but she just hadn't got around to it. Yes, Anne told me, she knew that doctors made the worst patients. But she felt that there was nothing wrong with her heart and she didn't need to waste her time. Her blood pressure was fine now; she just had a few palpitations that she was pretty much able to ignore.

So, I asked myself, if her heart was really of no concern, why was she talking about it? "I don't have time to see any doctors," Anne continued, the words tumbling out. Her life was stuffed to the brim with work, she said, her long days spilling over into weekends spent at the office where she was in charge of a group of radiologists. I wondered, too, how she had made time to see me—and why she'd really come. Anne looked lost, and behind her eyes seemed to be a distant sadness, a kind of longing for something she couldn't find. My own right mode was filled with a vague sense of pain, but at this point I couldn't place it, couldn't name it, so I just noted these internal sensations and filed them away in my mind.

Anne then told me that even with her professional success, she didn't feel very accomplished, and that her life was empty. There wasn't much else besides work. She had divorced her husband six years ago because "they just didn't have much in common." She hadn't been interested in dating when the twins were younger (besides, she was too busy), and she wasn't in a current relationship. Her daughters divided their time between her house and her ex-husband's in a nearby neighborhood. When I asked her about her relationship with the girls, she told me that they were "miniteens" who "didn't really want to bother with their parents." They were "very independent," she added proudly. Anne paused for nearly a minute, and I waited to see what else she would say. Then she looked at me with a puzzled expression and said, "Well, I'm here anyway . . . and I guess there has to be something more to life than just this." I took that to be a request for therapy.

When I asked Anne to tell me something about her upbringing, this is the story she told me:

When she was three years old, Anne's mother died of lung cancer and her father became very depressed. She was sent to live with her mother's parents in a nearby town, and she didn't see her father again for almost a year. During that time her father had been hospitalized, and when he was released he returned to live with Anne and her grandparents. When I asked Anne about that year, she said, "They were caring people, warm and loving," and then she paused for a few moments. "But it didn't last long," she added. "I was young, and my father came back, and, well, it all changed after that."

Anne's father remarried when she was five years old, and the new family moved across the country to settle in the Pacific Northwest, near Seattle. She didn't see her grandparents again until she was in college. Anne's father and stepmother had two more children, active boys born a year and a half apart, whom they doted on. Anne said she loved her brothers but felt ignored by her father. As for her stepmother, Louisa, she was "a robot of a woman" and a harsh disciplinarian who criticized Anne relentlessly. Anne's father never intervened.

One day when she was eleven years old, Anne had a particularly painful dressing-down from Louisa. Later, as she told me, she went for a long walk in the apple orchard in back of their house. She remembered making a decision: She promised herself that she would "never feel anything again." As she told me this, her face grew even more vacant, and she drew her index finger straight across the front of her throat. It was the gesture most people would recognize as "it's over" or "off with his head." But I wasn't at all sure Anne even knew she had made it.

"It worked. They could never touch me again. I mean, they didn't hurt me physically or sexually abuse me, but I never let them make me feel bad, no matter what they came up with. He and my stepmum just became nonpeople in my life. I ignored them from then on. I worked like crazy in school. My teachers loved me, and that was that. After college and medical school, I knew I would be okay. I think in many ways it all helped me become the successful doctor I am today. I suppose I should thank them . . . all of them . . . but I don't speak with them anymore. They wouldn't know what to do even if I did, I mean, to say I'm sorry, if they could. That's it. That's my story."

The session was over. Anne agreed to return, and then she went out into the rain.

KEEPING THE BODY OUT OF MIND

Halfway through Anne's second visit, a quotation from James Joyce that I'd heard somewhere popped into my head: Mr. Duffy "lived at a little distance from his body." It was in the way she moved, the stiffness of her gait, the way she held her hands motionless in her lap. (Her throat-cutting gesture stood out even more in retrospect.) It was also emerging from her account of a limited, rigid inner life lived only above the shoulders.

Anne told me she'd been quite artistic as a child—she'd excelled at drawing and loved to paint—although she'd had "no time for such things" in years. Unlike my patient Stuart, she did not seem to have a deficit in right-mode development; this was suggested by her artistic abilities and the fact that during her recounting of her personal history it was clear that she was aware of and able to articulate autobiographical memories in great detail, a specialty of the right brain. Moreover, sitting with me in the office, she expressed herself well nonverbally, making good eye contact and varying her facial expressions and tone of voice as different issues came up, which are other signs of right-mode development. Her left mode had also shown early strength; she'd been at ease with science and loved to solve math problems when she was in school. Her success as a radiologist supported my impression of at least some degree of horizontal integration; her profession required combining the spatial pattern recognition of the right mode with the analytic clinical mode of the left.

In our initial interview, Anne had spoken only briefly about her reaction to her mother's death: "She died, I was young, and I don't know what I would do without her." This confusion of past— "I was"—and present tense—"I don't"—is a window into possible issues of unresolved grief. I thought about how her mother's illness must have affected their relationship even before she died—how confused and frightened a toddler would be by her mother's inability to care for her. She had also experienced the sudden loss of her father, who disappeared and later returned, only to remain distant; and then she was taken away from the grandparents who'd cared for her lovingly for two years.

Next there was Anne's "decision" as an eleven-year-old "never to feel anything again." Anne spoke of this as a turning point in her young life. As I asked questions about her current experiences, the cutoff from her body became clearer. Anne "ate to live" and took little pleasure from food. She said matter-of-factly that she'd "never been a particularly sexual person." She'd never been involved in sports, and she had no physical fitness program.

The disconnect from her body wasn't complete, however. There was the matter of her palpitations. I asked Anne about their quality, frequency, and intensity, and she was able to tell me that they happened a couple of times a week, were "only mild" but—in contrast— "unnerving" enough to make her stop whatever she was doing.

She couldn't pinpoint anything that caused them. When I asked if she could sense her heart when it was beating normally, she said that she could not. But these sudden onsets of rapid, sometimes pounding, and irregular heartbeats "bothered" her. I urged her to go back to her doctor to be sure there was nothing to be concerned about. She said that she'd "think about it." Anne was an expert observer of interior anatomy in all its subtleties, but she refused to pay attention to her own body.

ESCAPE FROM PAIN

Anne had adapted to a painful situation by shutting off awareness of her feelings. What's wrong with that, you might ask? If our adaptations allow us to survive, why challenge them? Here's the basic problem: The conditions Anne experienced as a child—the painful loss of her mother and grandparents, her new family's neglect and harshness—no longer existed. She had adapted as best she could, but she'd had no support to help her resolve her losses—then or now. So her adaptation, which initially gave her strength and enabled her to move forward in her life, actually had come to imprison her. It kept her from being able to thrive.

Anne's decision to "never feel anything again" had effectively shut off the body proper from the neck down. It was as if she were trying to take refuge in her cortex, to cut herself off from the ongoing pain of criticism and isolation and unfairness. This adaptation may also have helped her leave behind—out of her awareness—her unresolved grief over her first great loss, her mother's death, which preceded all the others. Like all emotions, such overwhelming feelings are created throughout the extended nervous system, in the body, brainstem, and limbic areas; they directly involve our cortical regions as well. But if we can find a way to block subcortical input, if we can keep it from traveling upward into our consciousness-creating cortex, voilà!—we've "eliminated" our feelings.

No one knows exactly how our mind uses the brain to defend us from pain, but two things we do know from repeated clinical experience. One is that people do this quite often. As you'll see throughout this book, these adaptations can take many forms, from avoiding our feelings momentarily when we are overwhelmed, to long-term shutoffs, or to shutdowns like Anne's. The second thing we know is that somehow we—that is, our minds—can modify neural firing patterns to create what we need. For example, when we need to place something in the front of our mind, to focus our attention, we activate aspects of the prefrontal cortex on either side of the brain. So we can propose that one possible way the mind uses the brain to block something from awareness is by literally dampening the neural passage of energy and information from the subcortical regions upward to the cortex, especially to the parts of the prefrontal region that mediate awareness.

Here's another thing we know for sure: When we block our awareness of feelings, they continue to affect us anyway. Research has shown repeatedly that even without conscious awareness, neural input from the internal world of body and emotion influences our reasoning and our decision making. Even facial expressions we're not aware of, even changes in heart rhythm we may not notice, directly affect how we feel and so how we perceive the world. In other words, you can run but you cannot hide.

Colleagues of mine at UCLA have recently demonstrated that the pain of social rejection is mediated in an area of the middle prefrontal cortex that also registers physical pain from a bodily injury. This area is called the anterior cingulate cortex (ACC) and it straddles the boundary between our thinking cortex and our feeling limbic regions. In addition to registering physical sensations from the body and feelings from our social interactions, it regulates the focus of our attention. Because it links body, emotion, attention, and social awareness, the ACC plays a key role in the resonance circuitry that lets us feel connected to others and to ourselves. (In fact, the more we can sense our own internal world, utilizing the ACC and related areas such as the insula discussed in the Minding the Brain section "Riding the Resonance Circuits," the more we can feel the internal world of someone else.)

These research findings give us a new way to think about Anne: Her young mind would have been as driven to obliterate the chronic pain of loss and rejection as she would have been to escape physical pain. If she could shut down the activation of her ACC, perhaps she could "eliminate" the awareness of her pain. Standing in the apple orchard, Anne had found a way to exclude that pain from her conscious experience. The problem is, you can't eliminate bad feelings and keep the good. If you block lower input from reaching the ACC and the insula, you've blocked the source of emotion from reaching awareness. The result was a deadened emotional life and a cutoff from the wisdom of the body. The insula and ACC also appear to work together to create an overall self-awareness—something that seemed to be impaired in Anne as well.

BRAINSTEM SIGNALS: PAY ATTENTION!
FIGHT, FLEE, OR FREEZE?

We gain access to the body's wisdom through interoception, which literally means "perceiving within." Try pausing for a moment right now and just become aware of the beating of your heart and the in-and-out of your breath. These basic physiological processes are regulated by the brainstem; the brainstem also helps regulate our cortex by influencing our alertness and directly shaping our states of mind. You can pick up brainstem signals at any time by becoming aware of shifts in your breathing and heart rate—and also by paying attention to arousal itself.

Think of times when you realize you're feeling drowsy. You are focusing on the brain's alertness, noticing your capacity to attend to information—a teacher's lecture, for example, or this book you are reading. Perhaps you've returned to the same paragraph several times without taking it in and you're ready to admit that you are not in a state of mind to continue reading. You then choose how to respond: Should you have a cup of coffee or splash cold water on your face to try to wake up, or should you just take a nap? This is one way you regulate your internal world—by being able to monitor and then modify energy and information flow, in this case, levels of brainstem arousal.

The brainstem also works with the limbic area and cortex to assess safety or danger. When our threat-assessment system tells us we're safe, we let go of tension in our bodies and our facial muscles relax: we become receptive, and the mind feels clear and calm. But with an assessment of danger, the brainstem (along with the limbic and middle prefrontal areas) activates a decision tree: If we think we can handle the situation, we enter the fight-or-flight state of alert. This in turn activates the sympathetic branch of the autonomic nervous system (ANS). Our heart begins to pound as the body readies for action. Adrenaline pours into our bloodstream and the stress hormone cortisol is released; our metabolism is prepared for the energy demands ahead.

On the other hand, if we believe we're helpless, that there's nothing we can do to save ourselves, we freeze or collapse. Researchers call this the "dorsal dive," referring to the portion of the parasympathetic branch of the ANS that has been activated. This response goes back to our earliest evolutionary ancestors, and it's thought to have real benefits for an animal that is cornered by a predator. Collapse simulates death, so an attacker that eats only live prey may lose interest. Blood pressure drops precipitously in a freeze state, which could also reduce blood loss from wounds. In any case, it makes the animal or person fall limply to the ground as they faint, which maintains precious blood flow to the head.

If you are vertically integrated, you can read what your body is telling you about your safety or danger, including signs far more subtle than running away or fainting. You may feel a certain tension when you're walking down the street and only then realize that someone is following you. Or you get a feeling that you just can't trust the person you're talking with. In everyday life, having access to subcortical energy and information is also essential to thinking. Being aware of these subcortical impulses enables you to know how you feel, alerts you to your needs, helps you prioritize your choices, and then moves you to make a decision. This is how "gut sensations" or "heartfelt feelings" help us live our lives fully.

Since Anne had little interoceptive awareness, these subtle signs of safety, danger, or threat were probably muted or missing from her awareness as well. But even without awareness, these threat states, these brainstem-mediated neural shifts, can directly influence our thinking, our reasoning, and our sense of vitality. Someone can be ready to fight, vigilant for danger, or depleted by a sense of helplessness without knowing why. I thought Anne's palpitations might be in some way related to internal stress states. If a subtle internal or external threat led to adrenaline and cortisol release, her heart would pound, which would capture her attention—but since she had little consciousness of her internal state, or of its causes, she wouldn't know why it was pounding.

LIMBIC LANGUAGE: "PRIMARY" VERSUS "CATEGORICAL"

I'd been struck repeatedly by how confused Anne seemed when I asked her basic questions about how she felt in a particular situation. The cutoff seemed to extend to her relationships. She'd told me outright that she had few friends and no connection with her family. Staying away from her family as a child—and now as an adult— seemed self-protective, but I was concerned about the rather distant way she talked about her own daughters. They were the same age she'd been when she banished feelings from her life, and I knew that however "independent" children that age sometimes act, they do indeed need their parents.

In her first session, Anne had told me that her life was empty. Yet her refrain of "too busy" also conveyed that it was full to the brim in some ways. What seemed to be missing was the sense of energy and engagement that can give even ordinary experience richness, depth, and meaning.

To open the channels of vertical integration in Anne, to bring the signals of her body, brainstem, and limbic areas up into her cortical awareness, I first needed to open the doors of "emotional communication" between us. But when we talk about emotional communication, what do we actually mean?

If we focus only on the easily named and universally recognized emotions—such as anger, fear, sadness, disgust, excitement, happiness, or shame—we can miss the real richness of our minds: the realm of what I call "primary emotion." Primary emotion is the subtle music of the mind, the ebb and flow of energy and information that we sense during the moment-to-moment shifts in our internal state throughout the day. Sometimes, against this constantly shifting, changing background, an event occurs that orients our attention and activates our arousal, and the intensity of our arousal creates within us an emotion such as anger or fear. Even though these universal (or "categorical") emotions are recognized worldwide, in every known human culture, they do not emerge as often as you might think.

Consider the course of a day. How often do you experience clear, unambiguous anger or fear? For most, it is rare. Yet your inner world is filled with subtly textured, constantly changing states—what I am calling "primary emotions"—that continually color your subjective sense of being alive.

Thinking about these primary and categorical emotional experiences opens a new window on how we connect with others—and with ourselves. Young children need attunement with caregivers to feel seen and safe in the world. As parents, we can attune not only to our child's outbursts of categorical emotion—such as sadness or fear—but also to primary emotional states such as being energized, alert, focused, sleepy, or subdued. Parents who wait for a categorical emotion to arise before they "connect emotionally" with a child are missing the majority of important opportunities to attune. Attunement with a child's primary emotions is available moment by moment, as we pay attention to whatever has captured her attention. We can also tune in to our child's internal world by noting her levels of arousal. Is she engaged or depleted, lively or subdued? Having this primary emotional attunement to our children helps them feel deeply connected to others; as we resonate with them, they feel part of a larger "we."

Learning to track internal states—to become aware of our primary emotions—is a refined skill that begins when we're children and continues throughout our lives. Sensing this internal flow of energy and information is the essence of mindsight. As we first learn to pay attention to this flow through the attention our caregivers pay to us, we enter the world of knowing the mind. But Anne was not given the opportunity to learn how to sense her internal world from a safe, secure place after losing her mother and her grandparents. She, like so many of us, had to find a way to cloud her mindsight lens so as *not* to see her inner world. She learned to live a life devoid of meaning.

THE FEELING OF MEANING

Meaning is literally shaped by the limbic regions' appraisal process—the continual and immediate sorting of experience into "relevant or irrelevant," "good or bad," "approach or avoid." This, along with input from our middle prefrontal cortex, helps create the meaning of events in the brain. Meaning has a feeling to it, and establishing vertical integration for Anne would allow her to become receptive to this textured sense of significance coming from her inner world.

The cortex, especially in the frontal areas, can create abstract representations without input from the direct experiences mediated by the subcortical areas of the extended nervous system. We can think of the word *flower* but never sense the flower's aroma. We can paint that flower on canvas, but never lose ourselves in its textures and colors. Even right-mode visuospatial images can be sterile when devoid of access to subcortical input. There are musical virtuosos who leave audiences cold, literary scholars who are unmoved by the poetry they write about, doctors who diagnose but cannot connect with their patients. Integration requires openness to allow the many layers of our inner world to enter our awareness without rigid restrictions.

Words themselves are abstract representations that emerge like islands from a sea of associated meanings. Take, for example, the word *daughter*. If I say "daughter" to a young woman who's just heard the news that she is pregnant, that word will initiate a cascade of associations and responses. All sorts of beliefs may emerge: Daughters are fun. Daughters fight with their mothers. Men prefer sons. Will the pregnancy bring all the joys of her own relationship with her mother—or the pains of disappointment and confusion? Washes of sensation may fill her mind until she feels overwhelmed, unclear, cloudy. Maybe having a daughter would not be so good; maybe she'd be a better mother to a son.

With the word *daughter*, all of the young woman's own developmental history may be activated and revisited, with a mixture of old and new emotions. Was she close to her mother? Did she find her own voice, or did her mother overpower her? Taking on her mother's perspective, she might wonder how her mother felt about having a girl. How did she respond to her daughter's adolescence? Were her responses supportive or hostile or perplexing as she as a young girl matured physically, transformed from teen to adult, became sexually active, left home? And now that she is joining this passage of women from one generation to the next, how will her mother respond to the news of her pregnancy?

The meaning of *daughter* includes all of this and more, including the emotional associations that might arise if the young woman were to happen upon a mother-daughter pair at the park who appeared to be in rapt connection, exhilarated by each other, their laughter contagious yet private.

Now think of what *mother* meant to Anne. How could she stay open to her cascading associations, beliefs, concepts, developmental issues, and emotions? These elements of meaning, the architecture beneath our wash of feelings, would naturally flood her mind, intrude into her relationships, dis-integrate her brain. What choice did Anne actually have? Could she say, "Oh, no problem—let me be aware of this pain of loss of my mother. Let me be aware of this intolerable humiliation from my stepmother." Not possible. And so Anne discovered a survival mechanism: She cut herself off from meaning in her life. But while this was useful as a defensive maneuver in her childhood, it had become a fence that imprisoned her, cutting her off not only from herself but from her own daughters. Anne felt nothing and she was stuck. She had "a meaningless life."

THE FENCE OF DEFENSE

When strong primary feelings emerge or a particular categorical emotion arises, we may respond with an ingrained, learned reaction that is rooted in our past. If you grew up in a family in which anger was expressed as destructive rage, for example, you might get incredibly anxious whenever anger is expressed. In response to that anxiety, you may have learned to feel helpless and confused, causing you to freeze; or you may have learned to be fearful of rage, causing you to burst into tears and flee the scene; or perhaps you learned an aggressive "fight" response, causing you to meet anger with your own anger. Fight, flight, freeze—these are all emotional reactions to, yes, your own emotional responses.

Beyond our learned reactions to ordinary emotional threats, we also have patterns of adaptation that help us cope with overwhelming situations and with our reactions to them. These patterns of adaptation are sometimes called "defenses" and they shape the matrix of our personality: how we experience our inner world and interact with others. Here is the outline of the common pathway of defenses that is now accepted by many psychologists: An emotional response arises ➔ creating a reaction of anxiety/fear ➔ which initiates a defense. This defensive reaction shuts down the emotion, or at least the awareness of it, which then lowers the anxiety/fear and allows us to continue to function. This is why defenses are not only useful—they are often essential.

Defenses come in many forms. We can rationalize intellectually about a situation, minimizing awareness of our feelings by moving away from the more feeling right-brain mode into the logical left. This was Stuart's strategy. We can attempt to ignore a situation, skewing our perception to see just the positive side of an experience, a kind of "selective neglect." Some simply call this optimism, and it is a time-honored, and sometimes even healthy, strategy. When you are surrounded by lemons, make lemonade. Some people deal with a painful feeling by "projecting" it onto others and then hating them for it. This primitive and destructive adaptation is called "projective identification," the strategy that says the best defense is a good offense.

Whatever the defense, the idea is the same: We build a fence around our awareness so that we don't feel the anxiety or fear associated with feeling our feelings. These are usually automatic strategies, patterns of reactivity adopted without conscious intention or even recognition, and certainly without free will or choice. Anne's orchard "decision" was in fact an unusually conscious, perceptive moment of self-reflection. It was only later that her intentional suppression was transformed into automatic repression. During her childhood, Anne had no way to soothe her profound internal distress and interpersonal pain and so she could not remain open to their meaning, and her adaptation was to just "go cortical." Once she had blocked vertical integration, the primary function of Anne's body was to transport her head.

ATTENDING TO THE BODY

Anne and I were now at our fourth session, and I was able to present to her a plan for therapy based on our initial period of assessment. As a doctor, she was intrigued by the notion that her adaptation at age eleven might have persisted as a neurological pattern in her brain. I also told her that I thought she had been through a lot in her early years, and that I thought I could help her deal with whatever that time meant for her.

Anne and I needed to go on a journey together to help her feel receptive, to be attuned to herself, so that she could open up her awareness in new ways. She was up for the task, not certain what any of this would involve but willing to commit to a few months of therapy to find out. That was a good place to start. I told her, as I'd told Jonathon, that we'd need time to alter her synapses so that she could unlearn her old patterns and create new ones. Awareness, I went on, was the "scalpel" we would use to resculpt her neural pathways. Anne was intrigued by that image and wanted to know more. Now I knew I had captured her attention—the first step in changing her mind, and her brain.

I didn't want to distract her with the details of how awareness might enhance neuroplasticity, but I had some recent research in mind. The nucleus basalis, part of a region adjacent to the brainstem, has neural projections that secrete the chemical acetylcholine throughout the cortex. Acetylcholine is a neuromodulator, and its presence enables any neurons that are activated at the same time to strengthen their connections to one another. One theory suggests that we can use focused awareness to stimulate the nucleus basalis to secrete acetylcholine, thus enhancing neuroplasticity and learning. If this is so, it helps to explain why paying close attention gives our minds the power to change our brains.

All I told Anne was that through the work we would do she'd discover for herself the power of attention. We went over the basic mindfulness-of-the-breath exercise and we practiced some walking meditation. As we've seen with Jonathon's experience, learning mindfulness techniques can strengthen the hub of the mind so that internal sensations, such as bodily signals or waves of emotion, can be experienced with more clarity and calmness. My hope for Anne was the same, that with practice she'd strengthen the very parts of her brain that could not yet permit her to feel her feelings. She was game for taking on these forms of practice, not only in the office but as a daily "mental training" regimen at home. A weekly therapy session, an hour at a time, wasn't enough to focus her attention in an intense way. She would need regular synaptic exercise between her sessions with me. Reinforcing new synaptic linkages requires repeated neural firing to stimulate neuronal activation and growth—to SNAG her brain. As with Stuart, we could use the focus of attention to stimulate the activity and growth of areas that had been underdeveloped in childhood. In Anne's case, these regions would be the important circuits of interoception and self-regulation—of sensing the inner world and regulating that world—that had not been given the opportunity to grow well in her youth.

At her next session, I suggested we do a body scan, like the one I did with Stuart, thinking it would help her to gently become aware of her body in a nonthreatening way. I asked her to close her eyes and to look inward. Anne was fine as the focus of awareness moved up from her feet to her legs and then to her hips. I felt cautious as we progressed to the pelvic area. Anne had told me she hadn't been abused sexually, but this is one point at which anxiety sometimes emerges strongly during a scan. Anne focused with no problems. We then moved to the abdomen and the back, and she was still fine.

But when we focused on the chest, she started to breathe rapidly. Her hands started to shake. She made fists and pushed down with her forearms on the arms of the chair as if she were trying to hold down some feeling. Then she opened her eyes wide and said that she had to stop—she was hyperventilating and looked terrified. Anne had jumped from rigidity straight into chaos.

I was concerned that Anne was having a panic attack. We stopped the exercise and continued our session with her eyes open, and her agitation gradually subsided. She said she didn't want to discuss the experience. She was "fine" now; she simply "didn't like the body scan." We'd wait till later, when she'd developed more of an internal reserve to deal with upsetting sensations, to come back to that important source of bodily information. While research suggests that focusing attention on the heart can trigger both physiological reactions and an awareness of intense emotions, the specific nature of the feelings it had awakened in Anne was not yet clear. As our work continued, I hoped we'd learn more.

BUILDING INNER RESOURCES

The direct body-scan approach had triggered so much anxiety that Anne panicked, so I needed to choose more gradual ways to introduce her to body awareness. I started our next session by asking her simply to notice the movement of her fingers as she slowly opened and closed her palms. "Just noticing that," I said, "let yourself be filled with how the hand appears, and how it feels." We repeated the walking meditation, too, letting her feel the sensations of her feet with her eyes open.

Next I suggested that we develop a "safe place" into which she could always retreat—an image in her mind that she could draw upon to soothe herself whenever uncomfortable feelings arose. At first Anne had trouble coming up with an image. I told her this could be something from memory—a special vacation spot, her favorite room at home—or it could be entirely imaginary, a place where she could imagine herself being peaceful and contented, or at least safe and secure. Anne finally recalled a cove at the beach near her medical school. "I used to go there just to be with the waves," she said. "The sound of the waves, how they moved in and out, the curve of the beach, the sunny skies—everything gave me a feeling of things being okay." I asked her to sit with the image of the cove for a while, soaking in the sights and sounds and sensations. Then I told her to just notice her body and asked how that felt. When she said, "It feels good," I went on. "Being aware of the body, just sense whatever arises in your experience." I wanted her to create a neural association between her mental image of a place of safety and her awareness of bodily sensation.

This technique is used in several schools of body-focused therapy, and it had an entirely different purpose than the imagery work I'd done with Stuart. By creating that connection, Anne was able to experience and articulate what her body was feeling. She told me her abdomen felt soft, her face relaxed. Then she said her breathing was easy. She could feel her heart and it was "calm and steady." In contrast to her reactive panic during the body scan, Anne was now experiencing a state of receptivity. We were harnessing her regulatory prefrontal areas to help monitor and manage her internal states.

Another receptivity-enhancing technique I use involves systematically tensing and releasing the individual muscle groups of the body, from feet to head, which helps create a state of relaxation. Still others involve bilateral stimulation, whether by listening to alternating sounds or gently tapping on the left and right sides of the body. Some researchers believe that this creates not only relaxation but also increased sensitivity to mental imagery. But Anne felt most comfortable with her image of the cove and with the breath-awareness exercise I'd taught her first. We continued to practice them to give her confidence that she could go from reactivity to receptivity by her own mental efforts.

I wanted to keep her experience in the body on the positive side, so next I suggested that we try an exercise with color that evokes different feeling states. I do this work with a set of eyeglasses that have lenses made in various colors. Color is a powerful emotional cue for many people, but in Anne's case I asked her to focus on sensations in the body itself. Again, this seemed like a safe way—for some patients it's even a playful one—to awaken her awareness of shifts in physical sensation. With the first pair of glasses—green—nothing happened. "I don't feel a thing . . . just the usual . . . just blank." But when she put on the second pair—these happened to be purple—she exclaimed, "Wait—this is weird!" Anne said she had a "tingling feeling right up here," pointing to her upper chest.

After that, Anne felt her body change with each new color. Red evoked energy in her limbs "like ants running up my arms"; blue a deflated feeling in her abdomen "like a hole"; yellow a sense of constriction in her throat. This was not a test—each person has a unique response. The point was simply to create contrasting sensations, so that Anne could begin to recognize internal shifts.

Anne's initial response was excitement at her newfound ability, and we spent a good part of the session with the glasses, just letting her experiment with this neutral approach and find words to describe her body's sensations. But when I suggested that we might return to the body scan next time, she became frightened and hesitant. "I don't want to get into a panic again," she said, bringing her hand protectively to her heart. "Those feelings are not right . . . I can't handle them."

I reminded her that she now had her safe place as a resource at any time, and I assured her that we would move slowly. The internal world of Anne's childhood had been beyond what she could tolerate—at that time. Now she might be surprised to find that she could learn to tolerate what had once been intolerable.

WIDENING THE WINDOW OF TOLERANCE

Personal change, both in therapy and in life, often depends on widening what I call a "window of tolerance." When that window is widened, we can maintain equilibrium in the face of stresses that would once have thrown us off kilter.

Think of the window as the band of arousal (of any kind) within which an individual can function well. This band can be narrow or wide. If an experience pushes us outside our window of tolerance, we may fall into rigidity and depression on the one hand, or into chaos on the other. A narrow window of tolerance can constrict our lives.

In our day-to-day experience, we have multiple windows of tolerance. And for each of us those windows are different, often specific to certain topics or certain emotional states. I may have a high tolerance for sadness, continuing to function fairly well even when I or those around me are in deep distress. But even a lesser degree of sadness—whether your own or others'—may cause you to fall apart. In contrast, anger may be relatively intolerable for me; a raised voice may be enough to send me right out my narrow window. But for you, anger may not be such a big deal; you see a blowup as a way to "clear the air" and move on. In general, our windows of tolerance determine how comfortable we feel with specific memories, issues, emotions, and bodily sensations. Within our window of tolerance we remain receptive; outside of it we become reactive.

By now you've probably noticed that the window of tolerance matches the river of integration, which I introduced in chapter 4. The more freely that river can flow and the farther apart its banks are, the better we can attain and maintain integration and coherence. But if that flow is constricted, we're constantly in danger of hitting the banks. In many cases our well-being depends on widening the window of tolerance so that we can hold the elements of our internal world in awareness—without being thrown into rigidity (depression, cutoffs, avoidance) or chaos (agitation, anxiety, rage). As we develop mindsight, our windows of tolerance widen and we can experience the fullness of our lives with more acceptance and clarity.

If we move through life without mindsight, we may keep narrowing our window of tolerance around a specific emotion or issue. Then we may find ourselves either bursting the boundaries of that window and jumping into the chaos of reactivity, or avoiding situations that trigger such ruptures, restricting our lives without knowing why, not giving ourselves the freedom to escape our rigidity and empowering ourselves to grow. To widen our window, to make ourselves more adaptive and at ease with a particular feeling or situation, we need to change the associations that are embedded in the neural networks themselves.

"STAY WITH THAT": THE HEALING POWER OF PRESENCE

The presence of a caring, trusted other person, one who is attuned to our internal world, is often the initial key to widening our windows of tolerance. Because Anne did not have such relationships in her later childhood, her tolerance for awareness of bodily sensations and primary emotions had narrowed. Cutting off access to her subcortical input was once a means to survive—but now it was restricting her life. If I could be present fully with Anne, if I could let my own internal world resonate with hers and remain open myself, I could help her track her sensations and uncover their meaning, widening her windows of tolerance.

Recall that the resonance circuits include mirror neurons that would enable Anne to resonate with my own reactions to her. My being present fully with Anne at moments of distress could help her mirror my own inner feelings of safety. Here is a key fact about relationships: The resonance circuitry not only allows us to "feel felt" and to connect with one another, but it also helps to regulate our internal state. (It is the middle prefrontal area at the top of the resonance circuitry that shapes our subcortical states.) In other words, the interpersonal resonance between Anne and me could help widen her window of tolerance, so she'd feel safe enough to feel her own feelings. This is how in the moment, face-to-face, we help one another grow, and initiate the long-term synaptic changes that help us even when we're apart. And by continuing with her internal reflection practices at home—the mindfulness-of-the-breath and the walking meditations—Anne could further reinforce these synaptic changes, transforming the way she communicated with her own body.

At the beginning of our next session, I once again invited Anne to return to the body scan that had triggered her panic. It had now been ten weeks since our first session; during this time she had been doing her home practices regularly, and she and I had developed a trusting and collaborative relationship. Exercises such as the safe-place imagery and the colored glasses had helped her observe her inner world in a more objective and accepting way. She had also received a clean bill of health from her doctor, who had rechecked her heart and found nothing of concern physiologically. Still, I moved into the body scan slowly, giving her plenty of time to immerse her awareness in the subtle sensations from her lower limbs, her hips, her abdomen.

When we came to her chest, her panic began to emerge. She grimaced and her left hand went to her chest. She opened her eyes and said we had to stop. I reminded her that whatever that sensation was, she always had her breath awareness and her internal safe place to return to. If she felt herself getting too close to the edge, she could shift into a focus on her safe-place imagery of the cove and watch the waves go in and out for a while. She closed her eyes, focused on her breathing, and her face slowly relaxed. She opened her eyes again, looking right into mine, and said, "Thanks."

I suggested she might take a few moments and just let this new sense of openness fill her. As her body seemed to settle into the chair, and I saw her hands relax and her face become more supple, I said that she might just notice how she could use the focus of her attention to calm her body, and her mind.

Anne said she was "ready to dive in" and we went back to the body scan. When she focused on her chest region, the panic again began to emerge, but this time she said she could now sense it from a "more distant place." She had learned that she could just stay with her sensations, and that not only could she be "okay," the sensations themselves would change and become less overwhelming.

That's the strange thing about panic—when we lean into it, it loosens its grip on us. The power of reflection allows us to approach, rather than withdraw, from whatever life brings us. And when we learn to "stay with" a feeling, to give it its time in awareness, then we discover that feelings—even very strong and threatening feelings—first arise and then dissipate, like waves breaking on the shore. Panic is just another feeling, a set of neural firings in our brain. Learning to stay open and present to it, or to any other distressing feeling, is not easy, but it is an essential step in moving through the fences of defense.

THE WISDOM OF THE BODY

What was revealed as Anne learned to confront and regulate her anxiety, as she widened her window of tolerance? What sensations, images, feelings, and thoughts were now free to emerge? As we returned to the body scan during that session, Anne felt a wave of coldness in her chest and tightness in her limbs. Again she said it was hard to breathe. She spent a few moments at the cove, following her breaths as if they were waves on the shore, she later told me, and then said that she could continue.

As she stayed with her inner experience, images of her father and stepmother appeared in her mind's eye. She felt frightened of their faces and wondered if this panic was a fear of their meanness, of how they had mistreated her. She again focused on her breath to ground her in the hub of her mind, that open and receptive state of her regulating and self-soothing prefrontal cortex.

Now Anne began to tremble, her face looked tense, and tears began to flow down her cheeks. "I see a picture, but it's not something I remember . . . it's something I've seen, something I have. It's the only picture I have, the only thing I have left. It's a picture of me and my mother." Anne opened her eyes and looked at me. "I have that picture buried in my closet somewhere—I haven't looked at it in years." She seemed relieved but exhausted. It was near the end of our time, and I asked Anne if she'd like to just take a few moments to sense her breath, to let her body relax and her mind appreciate all that she'd been through during this session.

To make sure we had thoroughly explored her heart-mediated feelings of distress, we returned to the body scan during our next session. Anne's initial sensations of panic shifted gradually during the scan. She now began to feel a heaviness in her chest and a tightening in her throat. Then tears filled her eyes. As her panic was allowed to take its natural course, unhindered by defensive reactions, it moved towards completion, dissipated, and revealed an emotion that had been far more hidden in Anne, a profound sense of sadness. Now the essence of staying present for Anne was to allow these sensations of loss and grief to unfold in their own time.

In a subsequent session we simply sat together as she let the image of her mother holding her—the one in the photograph she'd remembered—fill her awareness. At first her tears were slow, a few drops she didn't seem to notice and did not wipe away. But as we stayed together with whatever she was feeling, she began to sob uncontrollably, her body bent over as she moaned in pain. I let her sense our connection with my own nonverbal signals—a sigh, a quiet "ummmm," the rhythm of our breathing in synch. When she opened her eyes and we looked at each other, I noticed my own tears.

"I know this sounds strange," she said, looking at me now with softer eyes than I'd ever seen in her, "but I can feel my mother's presence; I know she is here somehow with me."

Then Anne told me she had had a dream the night before our session. "I haven't had a dream in decades," she said, "and this was a strange one." Dreams are the work of sleep, one of the important ways we integrate memory and emotion. They occur when cortical inhibition is released enough to allow our subcortical limbic and brainstem regions to have a heyday with imagination and feeling. The dream itself is an amalgam of memories in search of resolution, leftover elements of the day's events, sensory inputs while we're asleep, and simply random images generated by our brain's wild activity during the rapid-eye-movement (REM) stages of sleep.

I thought it was a great sign that now, finally, Anne's subcortical regions were sending their input into her dreaming brain—enough for her to remember these internal images when she awoke. I listened closely.

"In the dream I am swimming to shore but the tide is going out and I can't fight it. Then my legs are tied to a boat that is headed out to sea, but I keep trying to get back. I'm pulling frantically with my arms, but I'm getting more and more exhausted. The boat just keeps moving and I can't see the shore anymore. I woke up this morning and felt panicky. It was awful."

I asked her to tell me more about what she'd felt when she awoke, and what came to mind now as she recounted the dream to me.

"I don't know. I think it's weird. Maybe I'm just too tired."

But a week later she described a second dream, and also the notes from the therapy journal she'd begun to keep. "I'm back in the water. Now I can see the shore. But the boat is moving again—I'm going quickly out to sea. I feel like I'm going to drown for sure. But then I reach down to my leg—I think I really did, it felt warm—and I pulled off the ropes. I freed my legs and started kicking like mad. Finally I got to the shore and collapsed in the warm sand. I remember just looking up at the sky, seeing the sun, and feeling safe. Then I woke up and knew it was all a dream, but I felt relieved."

This time she was more ready to talk about what these images might mean for her, and we explored her feeling of helplessness as she was pulled away from everything that was warm and solid in her life, and then her relief as she finally reached the shore again.

IMAGES OF HEALING

At the beginning of our next session, Anne handed me a large envelope. She had found the picture of her with her mother, which had been taken when Anne was about two years old. She told me that after her father remarried, he had destroyed anything that reminded him of her mother, and had never spoken about her. It was only after she'd left for college that she was finally able to visit her mother's parents, who gave her this photograph.

But in the envelope there were two pictures, an old snapshot and a larger print of the same image. Anne had scanned the old photo into her computer and then deleted the figure of her father, who had been "lurking" in the background. "I want to hold on to the part of memory that gives me warmth," she said. "I don't need to be tied to my father's mean wife, or to his grief."

The enlargement centered on little Anne and her mother, nestling together in an old-fashioned wing chair. Anne was in her mother's lap, pointing excitedly toward the camera with her right hand, while her left hand held on to her mother's encircling arms. Her mother was gazing down at her and smiling. It was a moment suspended in time, the child secure in her mother's embrace yet eager to reach out, the mother delighting in her daughter.

As I handed the photos back to her, Anne said, "I can see a certain sadness in her eyes." Her mother's cancer had been discovered when Anne was about a year and a half old. "I can only imagine how horrible that was for her, knowing she wouldn't be able to care for me, or see me grow up." We sat together, just staying with that feeling of clarity.

In the weeks ahead, Anne would also come to reflect on how difficult it must have been for her father—her grandparents had told her how much he had loved her mother, and how he'd fallen apart when she died. "I guess he did the best he could after she was gone," she said to me one day. "He was so young himself, only twenty-six. But I still can't understand why he just disappeared—and why he chose such a monster for his new wife. When my mum died, in many ways so did my dad."

Anne's grief was finally taking its natural course as she opened to all of her feelings—love, loss, confusion, anger, and even forgiveness.

Anne decided to stay in therapy beyond the few months she had committed to originally. As her work continued, Anne's life began to have a sense of vitality that had been missing for decades. She began to take time to exercise regularly. Her palpitations gradually reduced in frequency, then ceased altogether. She started to see some of her colleagues socially, outside the office. She also found the time to "just be with" her daughters, and she discovered that there were things they enjoyed doing together (it turned out that the girls liked art projects, too). Instead of catching up at the office every weekend, she made it a point to plan outings with them. "I know they won't be around much longer," she told me.

Anne feels more present in the room now. She holds herself differently; she seems at home in her body, her movements more fluid and relaxed. She has started to wear her hair down, flowing over her shoulders. And she has told me that she no longer feels empty inside.

8
Prisoners of the Past
Memory, Trauma, and Recovery

I WAS WITH BRUCE, waiting for a skirmish with the enemy, and thankful that he saw me as a friend, not a foe. The green and brown paint plastered on his face made him look more like a four-year-old at play than a thirty-four-year-old veteran, but the terror in his eyes and the power in his two-hundred-pound, six-foot-four frame made our situation all too real.

Bruce was one of the many men who had returned from Vietnam scarred from the inside out. Our paths came together under a bed at the Brentwood veterans hospital in Los Angeles, where he had been hospitalized for PTSD, post-traumatic stress disorder, a condition that had been given this name only a few years earlier. I was a new psychiatric trainee and Bruce was one of my first patients. Nothing had prepared me for the moment when he grabbed my ankles, pulled me into his "cave," thrust a broomstick into my hand, and shouted, "Shoot them if they come to get us."

There was no question that Bruce was lost in something akin to his imagination. But this wasn't a four-year-old's play; it felt to me as it was some sort of memory gone wild, some piece of his past that was quite alive and present in his mind, terrifying him—and now me. He peered into the room for what seemed an entire season of tropical monsoons, sometimes spotting the enemy coming towards us, then driving them away with his broomstick. He was thankful for my help, he said; he thought we made a good team.

After an hour of terror and anxious vigilance, Bruce's grip on his stick finally loosened. His tight, harsh voice went silent, his face softened, and then he began to whimper softly. I helped him out from under the bed and found a safe haven for him under the covers. I sat by his side until he fell asleep.

Shaken and confused, I wandered into the nurses' station and told the team nurse what had happened. "Oh yeah," she said, "those are just Bruce's flashbacks." She was trying to be helpful.

Later that day I had an hour of scheduled supervision and I asked my professor what in the world a "flashback" was. "A kind of memory from the past that continues to haunt the person in the present," he said. "We really don't know how a flashback happens." That's all he could offer—it's all anyone knew at the time—but it left me feeling restless and uncertain. I needed to know more.

I'd learned about our ability to selectively focus our attention on an imaginary world and suspend our critical judgment in order to enter it fully. Some call this "normal dissociation," a kind of willing suspension of disbelief, of getting lost in imagination. It's typical of kids at play, and we all do it when we get lost in a book or movie, become absorbed in a memory, or immerse ourselves in music. We narrow the spotlight of attention to one segment of experience, shutting out awareness of other activities of the mind. Of course, in everyday life, we can leave that absorbed state and come to dinner if we hear the call. But this was different.

Bruce didn't seem to experience that hour under the bed as something he was remembering, but rather as something that was actually happening in the present. He could also incorporate new items—the broomsticks, the cave under the bed, me—into the experience. This was more than becoming lost in a memory, or in imagination. Long-ago feelings, sights, sounds, and behaviors had become alive in his mind and interwoven with his moment-to-moment experience. To me it was clearly a memory, but for him it had lost—or perhaps never had—some labeling in his mind that identified it as a recollection. Instead those recollections seemed to be raw mental data, puzzle pieces of the past that had painfully exploded into his perceptions of the here and now.

Until recently, when we've been able to peer into the functions of the brain, we could only make guesses as to the mechanisms of memory, and the intimate ways our minds create our experience of reality. At the time I met Bruce, these intrusive memories were just the twists of the screw that tightened the clamps around his tormented mind. His recurring flashbacks were breaking him apart.

The next week I was told that someone had uncovered live grenades under a bush near the entrance to the building where we had holed up under the bed. Bruce denied knowing about them, but he was transferred to the locked ward after the staff discovered grenade pins in the locker in his room. A short time later he was transferred to another hospital and I never had a chance to work with him again. I still wonder into what distortion of memory those grenades might have fit.

HOW MEMORIES FORM AND REAPPEAR

In the years since my encounter with Bruce, a great deal of research has given us a framework for understanding and treating PTSD. By the late 1980s, a number of research centers had contributed pieces to the larger puzzle of how memory works in the brain. Those early findings have helped to build the interpersonal neurobiology view of trauma and trauma treatment that I'll present to you here. It may be too late to help Bruce, but we now have tens of thousands of soldiers returning from new wars whose minds urgently need healing. And there are even more people whose unintegrated trauma intrudes into their daily activities and relationships, overwhelms their ability to cope, and limits their lives—often without their conscious understanding of what is happening. Allison, whom you'll meet later in this chapter, was one of them. She was one of my first long-term patients, and her treatment brought me closer to seeing how trauma can fragment a life, and how it can be resolved.

To understand traumatic memories, it helps to go back to the basics of what memory is and how it is embedded in the brain. Memory is the way an experience at one time influences us at a future time.

As I discussed in "Neuroplasticity in a Nutshell" (page 38), experience for the brain means neural firing. When we have an "experience," clusters of neurons are activated to send electrical signals down their long lengths. The gene activation and protein production triggered by neural firing can create new synapses, strengthen existing ones, alter the packets of neurotransmitters that are released or the receptors that receive their messages, and even stimulate the growth of new neurons. It can also thicken the insulating myelin sheath around connecting fibers, increasing the speed of electrical transmission.

Neurons that fire together, wire together. In memory terminology, an experience becomes "encoded" by the firing of neurons in groups. The more often these neural clusters, or "neural net profiles," fire, the more likely they are to fire together in the future. The trigger that cues the retrieval of a memory can be an internal event—a thought or a feeling—or an external event that the brain associates in some way to a happening in the past. The brain acts as an "anticipation machine" that continually prepares itself for the future based on what has happened in the past. Memories shape our current perceptions by creating a filter through which we automatically anticipate what will happen next. In this way the patterns we encode in memory actually bias our ongoing perceptions and change the way we interact with the world.

Here's a key fact about memory retrieval that has been understood in detail scientifically only for the past twenty-five years: When we retrieve an encoded memory from storage, it does not necessarily enter our awareness as something coming from the past. Take, for example, your memory for riding a bicycle. When you get on your bike, you just ride—you fire off clusters of neurons that let you pedal, balance, and brake. That is one kind of memory: An event in the past (learning to ride) has influenced your behavior in the present (riding the bike), but riding the bike today doesn't feel like a memory of the day you learned to ride.

If, on the other hand, I asked you to recall the first time you rode a bicycle, you might pause for a moment, scan your memory storage, and perhaps come up with an image of your father or big sister running beside you, the fear and pain you felt when you fell, or the exhilaration when you made it to the corner. When these retrieval profiles fill your awareness, you know that you are recalling something from the past. This is also memory—but it is different from the memory that enables you to ride your bike.

These two kinds of memory processing are interwoven in the normal course of daily life. The kind of memory that enables us to ride the bike is called *implicit memory;* our ability to recall the day we were taught to ride is *explicit memory.* I am stressing this distinction because in everyday language we use the term *memory* to refer to what is technically explicit memory. But recent discoveries in the field of brain science allow us to understand the difference between implicit and explicit memory, as well as to grasp how implicit memory can influence our present without our awareness that something from the past is affecting us. It is these discoveries that have finally offered us an understanding of how Bruce's flashbacks may have developed.

Let's start at the beginning, with the implicit memories we lay down even before we are born.

IMPLICIT MEMORY: THE BASIC PUZZLE PIECES OF MENTAL EXPERIENCE

When my wife was pregnant with each of our two children, I used to sing to them in the womb. It was an old Russian song that my grandmother had sung to me, a child's song about her love for life and for her mother—"May there always be sunshine, may there always be good times, may there always be Mama, and may there always be me." I sang it—in Russian and in English—during the last trimester of pregnancy when I knew the auditory system was wired up enough to register sound coming through the amniotic fluid. Then in the first week after each child was born, I invited a colleague over for a "research study." (I know, it wasn't controlled, but it was fun.) Without revealing the prenatal song, I sang three different songs in turn. No doubt about it—when the babies heard the familiar song, their eyes opened wider and they became more alert, so that my colleague could easily identify the change in their attention level. A perceptual memory had been encoded. (Now my kids won't let me sing; I probably sounded better underwater.)

We encode implicit memory throughout our lives, and in the first eighteen months many researchers believe we encode only implicitly. An infant encodes the smells and tastes and sounds of home and parents, the sensations in the belly when she's hungry, the bliss of warm milk, the terror of loud and angry voices, the way her mother's body stiffens in response to a certain relative's arrival. Implicit memory encodes our perceptions, our emotions, our bodily sensations, and, as we get older, such behaviors as learning to crawl or walk or talk or ride a bike.

Implicit memory also harnesses the brain's capacity to generalize from experience, which is how we construct mental models from repeated events. This is one step beyond associations of neurons that fire together. The brain summarizes and combines similar events into one prototypical representation known as a schema. If a little boy's mother hugs him every evening when she comes home from work, he'll have a model in his mind that his mother's return will be filled with affection and connection.

Finally, implicit memory creates something called "priming," in which the brain readies itself to respond in a certain fashion. When his mother arrives home, the boy anticipates a hug. Not only is his internal world primed for perceiving that loving gesture, he'll move his arms in anticipation when he hears her car in the driveway. As we get older, priming continues to operate with more complex behaviors. If you've learned to swim, when you get your bathing suit on your behavioral repertoire for swimming is primed and readied to engage when you jump in the pool.

These six domains of implicit memory—perception, emotion, bodily sensation, behavior, mental models, and priming—are like the basic puzzle pieces of the mind that form the foundation for how the past continues to influence us in the present. After an experience is "over" and we move ahead down the river of time, what remains are these synaptic linkages that shape and filter our present experiences and sensations. Drawing on these implicit elements from the past, the brain—our associational organ and anticipation machine—continually readies us for the future.

Here are the three unique features of implicit memory: 1) You don't need to use focal, conscious attention for the creation—the encoding—of implicit memory; 2) When an implicit memory emerges from storage, you do not have the sensation that something is being recalled from the past; and 3) Implicit memory does not require participation of a part of the brain called the hippocampus. Going more deeply into each of these features will lead us into the mystery of Bruce's flashbacks.

ENCODING WITHOUT AWARENESS

If you had been a volunteer in one of the classic studies of divided attention, it would have gone something like this: The researcher gives you a set of headphones that play a different soundtrack into each ear and asks you to pay attention to the left side only. After a minute, she asks what you heard. Someone reciting a list of zoo animals, you say. What gender was the voice? Male, you reply. Fine. And what did you hear in the right ear? Just some vague mumblings, you say. And could you tell if it was a male or female voice? No, not even that.

But then the researcher administers what is called an indirect memory test, which reveals that the information from the right earpiece did indeed enter your mind and influence your memory—your implicit memory. You cannot recall that your right ear picked up a woman's voice reading flower names. But if you are given a set of partial word cues, such as "r _ _ e," you are more likely to fill in the blanks with the letters *o* and *s* to create *rose* than any other set of letters, even though you don't know why. If you had heard a list of food items, it might have just "come to you" to write in *i* and *c* for *rice*. This is priming at work in your language centers.

When your unattended right ear took in the data, your brain registered it in a form of perceptual implicit memory. It does this without passing the information through the hippocampus, the sea horse–shaped cluster of neurons in the limbic region that integrates widely separated areas of the brain. Direct attention harnesses the hippocampus; indirect attention—attention that does not involve your focal, conscious attention—encodes the memory without hippocampal involvement.

Again, an implicit-only memory is experienced in consciousness but is not "tagged" or felt as something emerging from the past. This is quite different from the idea of "unconscious memory," which implies something buried, inaccessible, or "repressed" and kept from everyday awareness. A reactivated implicit memory is fully conscious; it just lacks the sensation of recall.

This peculiar qualitative experience of implicit memory can be hard to grasp even for neurology students. So let me share a time-honored story about a nineteenth-century neurologist named Clafard and his unfortunate female patient. It seems that Madame X, the patient, could chat about everyday events with her doctor, but if he left the room and returned a few minutes later, she would not recognize him or remember their conversation. He would have to reintroduce himself formally and begin again. One day, Dr. Clafard hid a pin in his hand, so that when he greeted Madame X and shook her hand, she received a sharp prick that caused her to cry out. At their next meeting, Dr. Clafard introduced himself as usual and then extended his hand. Madame X pulled back and refused to shake it. When asked why, she replied, "Sometimes doctors do things that hurt you."

Here is a mental model based on implicit memory: "Sometimes doctors do things that hurt you." It presents itself as a fully conscious belief, but its origins in the past were not accessible to Madame X's awareness.

The implicit mental models that each of us has filter our ongoing perceptions and prejudge our experiences. And yes, they likely contribute to all sorts of attitudes and beliefs we carry around—whether about ourselves or other people. Our implicit models can manifest as a feeling in our bodies, an emotional reaction, a perceptual bias in our mind's eye, or a behavioral pattern of response. We do not realize we are being biased by the past; we may feel with conviction that our beliefs and reactions are based on our present good judgment.

If, for example, your parents ignored you when you came home all excited about being on the football team at school, that sensation of disapproval might generalize to other sports and then return when your own children become interested in athletics. Or perhaps your parents conscientiously avoided overt negative comments about people of other races, religion, or sexual orientation. But you still might have picked up nonverbal signs of irritation, distress, or disgust if you brought home a friend of a different background.

While these implicit mental models exist in all of us, with mind-sight we can begin to free ourselves from the powerful and insidious ways they create our here-and-now perceptions and beliefs. Seeing deeply and clearly into the inner world also gives us the opportunity to focus our awareness in a way that promotes the integration of memory. When memory is integrated, these separated implicit puzzle pieces of the past are linked together into the more complex—and flexible and adaptive—form of explicit memory.

EXPLICIT MEMORY: ASSEMBLING THE
PUZZLE PIECES OF THE MIND

Explicit memory begins to emerge and become observable by the second birthday, and while preschoolers may have quite vivid memories, most adults don't recall much about events before age five or six. (This phenomenon is called "childhood amnesia.") Explicit encoding depends on the ability to focus attention and integrate elements of an experience into factual or autobiographical representations. This allows us to create a scaffold of knowledge about the world, others, and ourselves that we can recall at will, reflect upon, and categorize in new and flexible ways. Parents instinctively strengthen this capacity in young children when they encourage them to talk about yesterday's trip to the zoo or who they saw at the playground that morning.

When we retrieve an explicit memory, we *do* have the feeling that we are bringing something from the past into our awareness. If I ask you to recall your last birthday, you may tell me where you were, what day of the week it was, and who came to celebrate with you. Your internal images are linked both to facts and to a sense of yourself within that particular experience or episode that took place in the past. These are the two forms of explicit memory: factual and *episodic* (remembering yourself in a single episode of your life), and as you may remember from Stuart's story, some people recall one form more easily than the other.

As life goes on, we accumulate episodic memories into larger files or clusters arranged along a time line. This clustered set of episodic memories is called autobiographical memory. Now you can tell a sad or funny story comparing your tenth birthday to your twentieth. You can construct a coherent narrative of your life.

With the maturation of the part of the brain that is required for explicit-memory encoding—the hippocampus—we are able to begin to create our factual and episodic memories. The hippocampus grows across the lifespan as it continues to lay down the explicit memory that enables us to know about the world and ourselves.

THE HIPPOCAMPUS: THE MASTER PUZZLE PIECE ASSEMBLER

If you recall our hand model of the brain, the hippocampus is located in the thumb area—the limbic region—on both sides of the brain. The left side works primarily with facts while the right specializes in self-related episodic memory. The hippocampus works closely with the other limbic areas, such as the fear-generating amygdala, to couple the details of an experience with the emotional tone and meaning of that event. It also has extensive linkages that enable it to combine previously separated neural firing patterns in the limbic region and throughout the cortex's perceptual and planning areas. In the left hemisphere it builds our factual and linguistic knowledge; in the right it organizes the building blocks of our life story according to time and topic. All this hippocampal work makes the "search engine" of memory retrieval more efficient. We can think of the hippocampus as a master puzzle piece assembler, which draws together the separate pieces of images and sensations of implicit memory into the assembled "pictures" of factual and autobiographical memory.

It requires focused attention to activate the hippocampus—to literally link together the neurally distributed puzzle pieces of implicit memory. When the images and sensations of experience remain in "implicit-only" form, when they have not been integrated by the hippocampus, they remain in unassembled neural disarray, they are not tagged as representations derived from the past, and they do not enter our life story as the unfolding narrative that explicitly defines who we are. Such implicit-only memories continue to shape the subjective feeling we have of our here-and-now realities, the sense of who we are moment to moment, but this influence is not accessible to our awareness. We have to assemble these implicit puzzle pieces into explicit form in order to be able to reflect on their impact on our lives.

WHEN THE HIPPOCAMPUS GOES OFF-LINE

Dr. Clafard's patient Madame X could not encode experiences into explicit memories because of a lesion near her hippocampus. I once met a man at a dinner party who had a similar condition. He politely told me that he had suffered bilateral hippocampal strokes and that when I returned from stepping away to get some water I shouldn't be insulted that he wouldn't remember who I was. Indeed, when I came back we started our introductions again.

But it doesn't take permanent, long-term damage to impair explicit memory. I once had a patient who told me the following story: He was about to take an overnight flight cross country, and he asked his doctor for something to help him sleep on the plane. The doctor gave him a new sleeping pill that had just come on the market, and my patient took double the prescribed dose, hoping to get a good night's rest. When he returned from the three-day trip, he had no (explicit) recall of anything following the first plane ride—even though the people he had met at his destination assured him he had seemed entirely awake and aware. (The pharmaceutical company that made the drug later halved the starting dose.)

Like some sleeping medications, alcohol is notorious for being able to shut off the hippocampus temporarily. Alcohol-based "blackouts" are not the same as fainting: The person is awake (though impaired) but does not encode experience into explicit form. People who have blackouts after drinking may not "remember" how they got home, or how they met the person who is in bed with them the next morning.

Rage can also shut off the hippocampus, and people with out-of-control anger may not be lying when they say they don't recall what they said or did in that altered state of mind.

Recent research suggests that other states of high emotion—beyond those we can normally tolerate—may also shut off the hippocampus by way of the high levels of stress they create. Excessive stress-hormone release in a state of terror, for example, may disrupt hippocampal integration.

When I first read this research, I realized I might finally be able to answer the question that had plagued me since I first met Bruce: What is a flashback? A flashback might be the result of the activation of an implicit-only memory of a traumatic experience. The perceptions, emotions, bodily sensations, and behaviors of a past time were fully in Bruce's awareness, but they were not tagged with the feeling that they were coming from the past. Because the hippocampus had been blocked, the raw moment-to-moment fragments of the experience remained as free-floating implicit puzzle pieces in disarray. The brain's circuitry that encodes experience into perceptions, sensations, and emotions remained active. But Bruce would not know that these internal images and sensations were coming from the past. His flashback would flood him with an "implicit-only" memory reactivation.

TRAUMA, MEMORY, AND THE BRAIN

Before the day when Bruce pulled me under the bed, we had started to explore his experiences in Vietnam. He began one session by saying that he didn't want to talk about that time, but he knew that he should. He was one of only a few survivors from his unit. As he spoke, his face became taut, his eyes seemed to roll upward, and his hands started to shake. Haltingly, in fractured pieces, the experience emerged—in words, in cries, in images Bruce could see and tried to describe, now with his hands raised, now covering his eyes, in shouts and in whispers that I can hear to this day.

Bruce's best friend from his hometown, Jake, was in his platoon. They were on patrol near the demilitarized zone when they were ambushed. Jake was hit in the head. Bruce, shot in the leg and unable to move, held his friend's limp body. Jake died in his arms as the rescue helicopters were coming in. With explosions all around him, Bruce just went blank. The next image he could assemble was the hospital ward in Saigon. The records showed that the medical team worried he had a brain injury; Bruce was unable to speak for weeks. Once he was returned to the States, Bruce tried to fit into civilian life.

His leg healed, but his mind was fractured. Ten years after his discharge from the military, he was admitted to the veterans hospital just before I began my training there.

What had happened in Bruce's brain? The best science can offer is a conceptual framework, supported but not yet proven by research. Under the stress of the extreme trauma he had undergone, Bruce had been filled with terror and collapsed from shock. Under such highly stressful conditions, the fight-flight-freeze response floods the body with the hormone cortisol, a chemical that has been shown to block hippocampal function. As I discussed above, anything that can temporarily shut down the hippocampus can also block the formation of explicit memories—an effect similar to that of alcohol or sleeping pills. This would have created the equivalent of a blackout, a chemically induced form of dissociation (but the chemical involved was cortisol, not drugs or alcohol). Paradoxically, the same intense reaction that led to the blackout and to the blockage of explicit memories would simultaneously heighten the encoding of implicit memory by way of the amygdala's release of another fight-flight-freeze chemical—adrenaline. High levels of adrenaline act to sear into implicit memory traces of the original traumatic experience— the feeling of terror, the perceptual details, the behavioral reactions characteristic of fight-flight-freeze, and any bodily sensations of pain that were suffered.

Here we see an explanation for the seemingly contradictory phenomena of PTSD—the fact that there is little or no explicit memory of the original event, yet the implicit memories that surface as flashbacks (or as other forms of free-floating implicit memory fragments) are incredibly vivid. By seeing how different kinds of memory depend on different regions of the brain we can finally make sense of the juxtaposition of enhanced implicit memory and blocked explicit memory that occurs during trauma.

Trauma may also shut down the hippocampus temporarily through the mechanism of dissociation. In the face of an overwhelming experience or threat to our survival, when there is no possible physical escape, not only do we release high levels of memory-blocking stress hormones, but the brain may find another form of escape by aiming the narrow channel of focal attention away from the threat. While we don't yet know exactly how this happens, awareness becomes completely absorbed in some nontraumatic aspect of the environment, or in the interior landscape of the imagination.

I don't know if dissociation was a part of Bruce's response to the ambush that killed his friend, but many people who have suffered trauma can clearly recall the dissociation that occurred during the event. Several years after I worked with Bruce, I treated a young woman who had been sexually assaulted at age fourteen when she was trapped by three men in an old storage shed. As the attack began, she told me, she noticed some small flowers poking through the broken siding at the corner of the floor. She focused on them, and they grew into an imaginary meadow in her mind. By remaining in that meadow, she averted her conscious awareness from her overwhelming sensations of pain and helplessness.

The problem with such a survival strategy is that while her immediate awareness was blocked, the temporary disabling of the hippocampus did not block implicit encoding of the experience. Fifteen years later, when she was taking a shower with her boyfriend, the sound of water on the floor of the shower suddenly triggered a full-scale flashback. It had been raining hard the day of the attack, and the implicit memories came flooding into her awareness as if she were being assaulted then and there—by her boyfriend. Fortunately, at the time she came to me for therapy I had by then incorporated into my work the latest findings about attention, the hippocampus, and memory, which meant I was able to make sense of her experience and knew what to do to help her.

While the hippocampus's role in the integration of memory was being revealed in scientific laboratories, it became apparent to me as a clinician that blocked integration might explain many common PTSD symptoms. Implicit-only memories and other cutoffs in the mind could be at the root of hyperarousal symptoms and explosive emotions; of numbing, disconnection from bodily sensations, and feelings of being "unreal"; and of various forms of reexperiencing the original trauma, including flashbacks and recurrent, distressing fragmentary recollections of the event while awake.

Sleep phenomena such as nightmares and REM disturbances are also key features of PTSD, and they offer us another window into the phenomenon of implicitly encoded traumatic memory fragments erupting into our lives years after the event with terrifying power. Before memories can be fully integrated into the cortex as part of permanent, explicit memory, they must go through a process called "consolidation," which seems to depend on the rapid-eye-movement (REM) phase of sleep. For many people with PTSD, REM sleep is interrupted, which may be a further explanation of why their traumatic memories remain implicit and are experienced as nightmares during sleep or reexperienced as symptoms while awake, such as those mentioned above.

For centuries, the intrusive, fragmenting, and numbing symptoms of trauma were recognized in soldiers and described in various ways, such as "shell shock." The PTSD diagnosis allows us to see the commonalities between battlefield experiences and the traumas that narrow and limit the lives of countless other people. I'd like to tell you about two patients who were among those others.

HARNESSING THE HIPPOCAMPUS TO HEAL TRAUMA

Allison first came to me at thirty-one because of persistent relationship problems, including significant sexual difficulties. When I asked her about her childhood, she told me that everything had been "fine," aside from her parents' divorce when she was three. Her mother had remarried when she was five and had two more children. After that, family life had been "normal." I wasn't sure what "normal" actually meant, but I thought we'd wait and see. There was plenty to explore in her present life.

A few months into therapy, she brought up a medical problem that had been plaguing her for some time. She had intermittent back pain, she told me, and now it was getting much worse. She was a fine arts teacher at a local school, and her pain was making it more and more difficult to work. She had consulted an orthopedist, who recommended surgery. But Allison thought she was too young to take such a drastic step, and she had read somewhere that back pain was often related to stress. She wanted to know what I thought.

I suggested that we try a body scan, moving upward from her feet, and that she just pay attention to her sensations. When we got to her back, she quickly became lost in terror. She recalled images of being at a neighbor's house one night, and then of their son's friend coming in drunk from a party and trying to have sex with her on the corner of the Ping-Pong table. He had jammed her back repeatedly into the table edge. As we explored these recollections over a series of sessions, it gradually became clear to her that the attacker was not the son's friend, but her own stepfather. With that realization, Allison's pain went away and never returned. She canceled her scheduled surgery.

I know you may think this is not possible, and if I hadn't been there myself or experienced similar therapeutic situations many times by now, I might feel the same way. And in fact, this was no "magic cure," because Allison's revelation was just the beginning of a great deal of hard work to rebuild her life.

Memory is not like a photocopy machine. When we retrieve memory, what we are recalling may not be accurate. Retrieval activates a neural net profile similar to, but not identical with, the one created at the time of encoding. Memories can indeed be distorted. We can have an accurate recollection of the gist—as when Allison remembered she had been assaulted—but the details may not be correct. In this case, over a long period of time Allison came to clarify details of her life narrative that were more horrific and painful than they first appeared.

Allison's memory retrieval was blocked for nearly sixteen years. Then what was initially recalled was distorted in a way that served to preserve the good image of someone important in Allison life: her stepfather. Many trauma victims struggle with these issues of accuracy. The reality is that memory is suggestible and many-layered. Fortunately, external corroborations are sometimes available to navigate these uncertain waters. Several months after Allison's pain went away, she attended a family reunion. There her younger half sister and half brother, whom she hadn't seen in two years, had the courage to tell her, when she asked them if they knew anything about what had happened at that party, that they had witnessed the assault. As witnesses, they too were victims of abuse.

You've probably noticed that Allison's initial distortion also protected an even more important person in her life: her mother.

Why hadn't Allison gone to her mother after her stepfather's attack? Even if she'd felt too ashamed to speak, why hadn't her mother picked up that something was wrong?

When families do not offer a place for children to express their feelings and recall what happened after an overwhelming event, their implicit-only memories remain in dis-integrated form and they have no way to make sense of their experience. As we discovered in our work together, Allison's family had become a zone of silence long before the night of the Ping-Pong table. Her stepfather had been intrusive in various ways almost from the time he married her mother. Her mother had sometimes turned a blind eye and sometimes even facilitated the abuse, in effect sacrificing Allison to her husband and her new family. This kind of early and repeated abuse, coupled with the lack of anyone to turn to for safety, is now known to underlie the development of dissociative disorders. At the very core of her self, Allison was unable to know explicitly what she knew implicitly all too well. She was unable to make sense of her life story.

Allison's therapy continued for many years, and here I can give you only the broad steps of our journey. Our goal was not only to integrate her disruptive memories, but also to help her navigate her current relationships and remain present in the face of the ongoing stresses in her life. Allison needed to build the skills of resilience and personal strength. After her betrayal by those closest to her, how could she learn to protect herself and at the same time learn to trust others?

In my own mind, I imagined the sequence of resolution going something like this: Attachment betrayals and traumatic experiences produce impairments to integration. In the domain of memory, this results in implicit puzzle pieces remaining in disintegrated form. These implicit-only pieces of the past intrude on the present, creating reexperiencing events (such as flashbacks and back pain), avoidance (without realizing why, Allison never played table tennis or pool), and numbing (which was at the core of her sexual problems). This fragmented experience needed first to be integrated into explicit memory and then incorporated into a much larger sense of who Allison was.

We would explore the unresolved memory representations, but with a dual focus of awareness. This means that while one focus of awareness is on the here and now, another is on the there and then. We would develop a set of resources and keep those readily accessible in the present moment, even as she and I moved into the second focus of awareness—the memories themselves, the implicit reactivations.

My job was to help Allison maintain a sense that we were together, that she was not lost in the past, even when her awareness was on the sensations of implicit memory. If she could move in and out of the past flexibly, and with less pain, she would feel safer. As background to the work we needed to do on her journey to integrate memory, I taught Allison about the brain, the mind, and memory, as I'd come to do with virtually all of my patients. I also taught her basic techniques such as breath awareness and helped her develop her safe-place imagery.

Allison's favorite technique was a variation on the wheel of awareness. I asked her to visualize a file cabinet in a locked room in an imagined house of her mind. In this file cabinet was locked the memory of whatever we were working on, especially when that memory felt intense and unresolved. She alone had the key to the room and could open the door. She alone had another key to open the file cabinet. At any time she could leave the room, shut the door, and cross the hall to another room, where she could watch the video of the event (this was before DVDs) on a playback machine. She could start, stop, freeze-frame, rewind, or fast-forward at her will. With this capacity to pull back before she became lost in the implicit world, she was prepared to dive into her sea of memory.

Brief immersions in the moment-to-moment sensations of her implicit memories seemed to be essential. Allison needed to be able to connect with and track those long-ago bodily experiences. But I didn't want her simply to "relive" the trauma. She needed to be aware simultaneously that she was *with me,* that she was safe, and that she could return at any time to the present and to all of her adult strengths and resources. My old memory mentor—one of my greatest teachers—had a powerful saying: "Memory retrieval is a memory modifier." In the presence of an attuned other, and with the help of tools such as the filing cabinet and her peaceful-place imagery, Allison could retrieve and make explicit her previously implicit-only memories. Unlike a flashback, which seems to engrain a dis-integrated state more deeply each time it occurs, this dual focus—on the memory, and on the self experiencing the memory (which we could call retrieval with reflection and release)—seemed to harness the hippocampus in a new way. It is now a dozen years later, and when I saw Allison recently, she told me her flashbacks had never returned.

However, finding a way to embrace the truth did much more than resolve Allison's symptoms. As she explored the many layers of her adaptations to the pain of her childhood, Allison wove her newly assembled explicit memories into a larger, more coherent framework for what made Allison Allison. She experienced a new sense of energy and pleasure in her life. She had recast herself not only as someone who had survived, but as a person who could thrive. This way of integrating memory seemed to empower Allison—and others since—to reclaim the authorship of her own story as it was being woven during the course of her hard and courageous internal work.

FALLING FLAT ON HER FACE

Even if we haven't suffered repeated or life-threatening trauma, implicit-only memories can become prisons that constrict our lives. One of the most striking examples of this is my patient Elaine.

Elaine was a twenty-six-year-old graduate student who came to me about her anxiety over finishing school. She told me right away that she feared she would "fall flat on her face" if she accepted the job offer she had already received during her final semester. Over the next few weeks, I tried several approaches to her fear of new challenges and her insecurities about the competitive job market. She was politely receptive to my ideas, but she remained stuck and frozen.

Something about the way she'd described her fear—"fall flat on my face"—had lodged in my mind, but I didn't know what to make of it. One day, when she was telling me yet again about her fears about finances and logistics, I suggested she just become aware of her body. She paused and then began to shake. She grabbed her arm and said "Ouch! What is going on?" I asked her to just stay with the sensation and see where it took her. The pain moved up her arm and into her jaw. She then grasped her mouth and started to cry. Soon she was describing what was going on in her mind. She was three years old and had fallen off her new tricycle. Then she remembered, explicitly, that the fall had broken her arm and fractured both of her front baby teeth. She and I were both startled by the intensity of her bodily sensations, which she experienced initially as "just pain" and not as a recollection.

Elaine's arm had healed, and the accident hadn't affected her adult teeth, but it did affect her adult mind. It had created an implicit mental model, or schema, coupling novelty and enthusiasm with intense fear and pain. She had applied this learned fear in her academic work, her employment, and even in her personal relationships. The message was "trying new things could result in disaster." She literally feared she'd "fall flat on her face" if she took the kind of job she had worked so hard to obtain.

As with Allison, I taught Elaine specific ways to remain present and safe in the face of her fears, and she gradually found a way to move towards her excitement at school or with friends. Once fully embraced and examined, her fear could be properly located in time, acknowledged for the frightened three-year-old's experience it was, and woven into a new story. Now that she was no longer the prisoner of an unexamined past, Elaine could take charge of her life with a new sense of vitality and freedom.

Working with patients like Elaine and Allison has convinced me that a dual focus is one of the crucial elements in trauma therapy. This simultaneity of conscious attention, in which you are focused both on the past and on your present-day self reexperiencing the past, is an active, engaged process that initiates hippocampal assembly of those strewn puzzle pieces of implicit-only memory. Elaine's observing self could witness herself sensing the images and bodily feelings from the past—but in the presence of a trusted other who could tolerate the painful memories. In this setting of emotional safety, the retrieved memory became less charged. Together we could identify her sensations as recollections, not as part of a new event, and she could then integrate the pieces of memory into a larger, more coherent sense of self. Once the hippocampus could perform its integrative functioning, her memories could take their place in an active and open life narrative, the unfolding story of who Elaine could become.

Unexamined implicit memories can shape our beliefs and our expectations. We may be tempted to see these embedded emotional responses as "intuitions" or "gut reactions" that give us deep insight into our ongoing experience. Like Elaine, we may also justify them rationally, coming up with numerous plausible reasons for our reactions. But such automatic responses may instead be leftover rubbish from painful moments of our unexamined past, not worthy of being trusted to guide our decisions or actions. They can make us irrational over crepes. And they can tie us to painful past events that we'd never intentionally choose to re-create.

But when we integrate those embedded experiences into our present consciousness and recognize them as implicit memories—not valid intuitions or reasoned decisions—then we begin to offer ourselves the means to become awakened and active authors of our own life story. And as we'll see in the next chapter, how we come to make sense of our lives is another crucial form of integration.

9

Making Sense of Our Lives
Attachment and the Storytelling Brain

MY COLLEAGUE REBECCA came to her postgraduate medical training after a hard-won battle with a history of abuse. She was the fifth of seven children born to an alcoholic mother and a father with bipolar illness, and her family life was filled with chaos and instability. She never knew what condition her mother would be in from day to day; her father, who refused mood-stabilizing medications, careened between mania and depression. When we were on call together late at night in the hospital where we worked, she'd tell me how her siblings and she would hide in the attic, where her oldest sister, Francine, would read them stories by flashlight while their mother raged downstairs. Francine would huddle with Rebecca, holding her and the others and pretending they were "camping out" during those emotional hurricanes. "Life was a nightmare," Rebecca said, "and we never really knew when we'd wake up."

Yet to me Rebecca seemed incredibly calm, notable for her ability to handle complex situations both with our psychiatric patients and with our fellow residents, one-on-one or in intense group discussions. One day I asked her: How did she make it through?

"It wasn't easy," I remember her telling me, "but besides my own sister, my mother's sister Debbie saved my life. She helped me see that I wasn't crazy. And even when I couldn't go to my aunt's house, she was always there for me. I knew I was inside her heart."

I will never forget that phrase: "inside her heart." Rebecca's feeling that she was closely held "inside" someone else made all the difference.

It wasn't until years later that I would come upon the research demonstrating how crucial it is to our development to have at least some relationships that are attuned, in which we feel we are held within another person's internal world, in their head and in their heart—relationships that help us thrive and give us resilience. And only later still did I learn how the neural networks around the heart and throughout the body are intimately interwoven with the resonance circuits in the brain—so that when we "feel felt" by another it also helps us to develop the internal strength of self-regulation, to become focused, thoughtful, and resourceful. Being close to someone early in our lives gives us the clarity to know how we feel, and the ability to feel close to others. Long before researchers began to unravel these neural mechanisms, poets and children like Rebecca knew that the heart is indeed a wise source of knowing.

PATTERNS OF ATTACHMENT

Rebecca felt that her heart connection with her aunt had saved her. But how? And how could she, in those late-night discussions, tell me about her painful past in such a clear and open manner?

For me, the explanation lies in some of the most exciting research done in psychology during the past thirty years: the ongoing exploration of early attachment. We have discovered that our early relationships shape not only how we narrate the stories of our lives when we reach adulthood, but also how our minds develop in infancy and childhood. First I'll delve into these fascinating findings as they affect children, and then I'll show you how I apply the findings as I treat my adult patients.

The initial stage of research was done with children during the first year of life. Trained observers made home visits throughout the year to assess mother-infant interaction on a standardized rating scale. Then, at the end of the year, each mother-infant pair was brought into the laboratory for a test that lasted approximately twenty minutes. This test is known as the "Infant Strange Situation," because it focuses on what happens when one-year-old babies are separated from their mothers and left in a "strange situation"— either with a stranger or alone. The idea is that separating a

one-year-old from his mother is inherently stressful and would activate the baby's attachment system—the way he has come to connect
with his primary caregiver. Researchers looked at how the children
reacted to the separation itself, and then how they responded when
their mothers returned.

These studies have now been done thousands of times by the
original researchers and replicated hundreds of times by scientists
throughout the world. How an infant responded to the Strange
Situation in the lab could be directly correlated with the careful and
repeated observations of the child and caregiver at home. The
reunion phase turned out to be the key. Researchers initially found
three basic patterns and then later delineated a fourth, based on how
the child greeted the mother when she returned after the separation,
how easily the child's distress was soothed, and how rapidly he
returned to play with the enticing toys in the room.

About two-thirds of children in the general population had what
was called a *secure* attachment. The child showed clear signs of missing the parent when she left, often by crying. He actively greeted
her—usually by seeking direct physical contact—when she returned.
But then he settled down quickly and returned to his childhood task
of exploration and play. Looking back at the home-visit observations, these were the children of parents who were sensitive to the
baby's bids for connection, who could read the baby's signals and
then effectively meet his needs.

Around 20 percent of the children demonstrated what was called
an *avoidant* attachment. They focused throughout on the toys or on
exploring the room, showed no signs of distress or anger when the
parent left, and ignored her or actively avoided her when she
returned. What do you think the child's experience during the first
year of life had been? As you may have guessed, home observations
showed that the parent did not respond to the child's signals in a reliable and sensitive manner, even ignoring these signals and seeming
to be indifferent to the child's distress. So the infant gradually
learned something like this: "Mother doesn't help me or soothe me,
so why should I care whether she goes or returns?" Behavioral avoidance is an adaptation to this kind of relationship. In order to cope,
the child minimizes activation of the attachment circuitry.

Another 10 to 15 percent of the children had what came to be called an *ambivalent* attachment. In this case, the first year of life was filled with parental inconsistency. Sometimes the parent might be attuned, sensitive, and responsive, at other times not. If you were an infant, how would you respond? Would seeking contact with this parent soothe your activated limbic distress? In the Strange Situation, the ambivalently attached infant often seems wary or distressed even before the separation. She seeks out the parent upon reunion, but is not readily soothed. She may continue to cry instead of returning quickly to the toys, or she may cling to the parent with a look of concern or desperation. Contact with this parent clearly does not give her a sense of relief, and there appears to be an overactivation of the attachment circuitry.

In later studies, a fourth category of attachment was added. Called *disorganized*, it appears in about 10 percent of the general population but in up to 80 percent of high-risk groups such as the children of drug addicted parents. It is quite upsetting to observe what happens when the parent returns. The infant may look terrified; he approaches the parent but then withdraws from her, freezes or falls down on the floor, or clings and cries while simultaneously pulling away. Disorganized attachment results when parents show a severe and terrifying lack of attunement, when they are frightening to their infants, and when they themselves are often frightened. The children in the other three patterns have developed organized strategies for dealing with a sensitive, disconnected, or inconsistent caregiver. But here the child cannot find any effective means to cope. His attachment strategies collapse.

How do these findings fit with our discussion of integration? You may have noticed that energy and information move in a harmonious flow in secure attachment; tend towards rigidity in avoidance; towards chaos in ambivalence; and alternate between rigidity and chaos in disorganization. Only in the disorganized form does the flow move beyond the "window of tolerance" that I introduced in chapter 7—resulting in a collapse in coping ability. In the remainder of the chapter, we'll see how these early patterns of behavior can persist as characteristics of the mind later on in life.

Many of the first children studied in the Infant Strange Situation have now been followed for more than a quarter of a century.

Despite all the other influences on their development during that time, their personal characteristics tended to diverge in predictable ways.

In general, the securely attached children were found to meet their intellectual potential, had good relationships with others, were respected by their peers, and could regulate their emotions well. Although the attachment researchers did not study the brain directly, these overall outcomes parallel our middle prefrontal functions in many ways: securely attached children developed good bodily regulation, attunement to others, emotional balance, response flexibility, fear modulation, empathy and insight, and moral awareness. (The ninth function, intuition, has not been studied yet.) From the viewpoint of interpersonal neurobiology, this strongly suggests that secure parent-child interactions promote the growth of the integrative fibers of the middle prefrontal region of the child's brain.

In contrast, those with an avoidant attachment to their primary caregiver tended to be restricted emotionally, and their peers often described them as aloof, controlling, and unlikable. Children whose primary attachment was ambivalent revealed a great deal of anxiety and insecurity. And those with disorganized attachment were significantly impaired in their ability to relate to others and to regulate their emotions. Further, many had symptoms of dissociation that placed them at heightened risk for developing PTSD after a traumatic event.

You may wonder—and I certainly did: Couldn't these differences be genetically based? Most of these parent-child pairs share at least half their genes, so perhaps the correlation between attachment patterns and later personality can't be attributed to anything the parent did—or failed to do. Studies have indeed established that the closer we are genetically, the more traits we share: from intelligence and temperament, to specific personality characteristics such as political orientation, cigarette smoking, and television watching. But one feature that stands apart from this list is attachment. And that's not just the opinion of the psychotherapy community: One of the leading researchers on the genetics of personality spontaneously made this statement at a national scientific meeting. Attachment patterns are one of the few dimensions of human life that appear to be largely independent of genetic influence. We can see this quite directly in those instances where a child has a distinct pattern of attachment to each of his caregivers. If attachment were genetically determined, how could the child's single set of genes allow for those differences? Furthermore, research with foster and adoptive children—who are genetically unrelated to their caregivers—has discovered these same patterns.

Naturally, who we become as adults is shaped by many factors—including genes, chance, and experience—in addition to our earliest attachments to our caregivers. But anyone who doubts the influence parents have on their children must deal with these extensive studies of attachment. They demonstrate clearly that what parents do matters enormously.

CREATING A COHERENT LIFE STORY

Why do we parent as we do? When researchers asked this question, they hypothesized—as many of us would—that it is the childhood experience of parents that predicts how they behave with their own children. This sounds plausible, but it turns out *not* to be quite right.

When I first heard about what the researchers actually found, it changed my life and my understanding of the life of the mind. The best predictor of a child's security of attachment is not what happened to his parents as children, but rather how his parents *made sense of* those childhood experiences. And it turns out that by simply asking certain kinds of autobiographical questions, we can discover how people have made sense of their past—how their minds have shaped their memories of the past to explain who they are in the present. The way we feel about the past, our understanding of why people behaved as they did, the impact of those events on our development into adulthood—these are all the stuff of our life stories. The answers people give to these fundamental questions also reveal how this internal narrative—the story they tell themselves—may be limiting them in the present and may also be causing them to pass down to their children the same painful legacy that marred their own early days. If, for example, your parent had a rough childhood and was unable to make sense of what happened, he or she would be likely to pass on that harshness to you—and you, in turn, would be at risk for passing it along to your children. Yet parents who had a tough time in childhood but did make sense of those experiences were found to have children who were securely attached to them. They had stopped handing down the family legacy of nonsecure attachment.

I was excited by these ideas, but I also had questions: What does "making sense" really mean? How can we accomplish it, and how does it occur in the brain?

The key to making sense is what the researchers came to call a "life narrative"—the way we put our story into words to convey it to another person. How an adult told his or her story turned out to be highly revealing. For example, people who were securely attached tended to acknowledge both positive and negative aspects of their family experiences, and they were able to show how these experiences related to their later development. They could give a *coherent* account of their past and how they came to be who they are as adults. In contrast, people who had challenging childhood experiences often had a life narrative that was *incoherent* in the various ways I'll describe in the following pages. The exceptions were people like Rebecca. Based on the facts of their early childhood, they would be expected to have an avoidant, ambivalent, or disorganized attachment as children and an incoherent life narrative as adults. But if they had a relationship with a person who was genuinely attuned to them—a relative, a neighbor, a teacher, a counselor—something about that connection helped them build an inner experience of wholeness or gave them the space to reflect on their lives in ways that helped them make sense of their journey. They had what the researchers called an "earned secure" life narrative. Such a secure narrative has a certain profile; we can describe its features. Even more important, like Rebecca we can change our lives by developing a "coherent" narrative even if we did not start out with one.

This is such a crucial point that I'll repeat it: When it comes to how our children will be attached to us, having difficult experiences early in life is less important than whether we've found a way to make sense of how those experiences have affected us. Making sense is a source of strength and resilience. In my twenty-five years as a therapist, I've also come to believe that making sense is essential to our well-being and happiness.

THE ADULT ATTACHMENT INTERVIEW

The research instrument that measures how we have "made sense" of our lives is called the Adult Attachment Interview, or AAI. If I were to give you a version of the AAI, I would ask you a series of questions paraphrased like these. What was your childhood like? What was your relationship like with each parent—and were there other people with whom you were close as a child? Whom were you closest to and why? I'd ask you to give me several words that described your early relationship with each parent or caregiver, and then I'd ask for a few memories that illustrated each of those words. The questions go on: What was it like when you were separated, upset, threatened, or fearful? Did you experience loss as a child—and if so, what was that like for you and for your family? How did your relationships change over time? Why do you think your caregivers behaved as they did? When you think back on all these questions, how do you think your earliest experiences have impacted your development as an adult? And if you have children I'd ask you these questions: How do you think these experiences have affected your parenting? What do you wish for your child in the future? And finally, when your child is twenty-five, what do you hope he or she will say are the most important things he or she learned from you? That's essentially it.

Answering this set of open-ended questions is like diving deeply into areas of untapped memory. When I was doing research with the AAI, many subjects told me that the interview was the most helpful therapy session they'd ever had. As a therapist I found this especially amazing, because the research protocols required that I be as neutral as humanly possible. Nevertheless, something about these questions repeatedly prompted new discoveries even in individuals who'd had years of therapy.

If I had administered the AAI to you in a research setting, your responses would have been recorded and then transcribed.

The transcription would then have been carefully analyzed by a researcher trained to code the results. We would be watching for how you presented the material, noting if the details of memory corresponded to the generalizations you offered, keeping track of the unfolding of your story to see if it made sense and held together logically, and observing patterns of response such as insisting you did not recall the past or getting confused between past and present. The transcription would also be assessed for how you monitored what you were saying as you went along, and how you took me into account as you spoke—did you try to make sure that I understood what you were saying? In this way, the "narrative analysis" becomes as much an evaluation of interpersonal communication as it is a study of your own internal process.

The AAI evaluation accepts that memory is fallible. As you've seen, memory is not a photocopy machine, and it's highly suggestible, readily conforming to others' expectations—and to our own. Even at our most honest moments we say things we think others expect to hear, and we say them in ways that make us appear as we want to appear. For these reasons the analysis does not presume the accuracy of the facts as stated. Instead it focuses on the coherence of the story.

This analysis of the responses to the AAI questions reveals the "adult state of mind with respect to attachment" that turned out to be so powerfully predictive of parenting behavior and of how the children of these adults responded in the Infant Strange Situation. Later studies also revealed that a child's attachment behavior in childhood predicted the type of narrative he or she developed as an adult. I'll be exploring these links in the remainder of the chapter, but here is a quick summary of the major categories.

THE CORRESPONDENCE OF ADULT AND CHILD ATTACHMENT

ADULT NARRATIVE	INFANT STRANGE SITUATION BEHAVIOR
SECURE	SECURE
DISMISSING	AVOIDANT
PREOCCUPIED	AMBIVALENT
UNRESOLVED/DISORGANIZED	DISORGANIZED/DISORIENTED

Although there is a cause-and-effect relationship here, as mentioned earlier the legacy passed down from parent to child is not necessarily destiny. As happened with Rebecca, one can eventually develop an "earned secure" narrative despite having an insecure attachment and suboptimal experiences in early childhood.

A NEW WINDOW ON THE MIND

It's now twenty years since I first pored over narrative analysis sheets to learn to conduct AAI research, but the AAI still plays an important role in my everyday psychotherapy work with patients. Today I listen to my patients' narratives for an understanding that goes far beyond any statistical category, and I have discovered that they illuminate many layers of the mind. The AAI questions help us explore how our childhood relationships shaped the patterns of our internal world, especially influencing our windows of tolerance and our capacity to reflect on our own internal world. Patients with coherent narratives have wider windows of tolerance and more robust mindsight skills. In other words, secure attachment seems to go hand in hand with integration. And if a patient is insecurely attached, the AAI helps me find ways we can work together to increase integration and create an "earned security."

In the following pages, I'll share patients' responses to the AAI questions and what they reveal about how they've made sense of their lives. I'll look at how the four categories of child attachment—secure, avoidant, ambivalent, and disorganized—turn up in adult narratives. I'll also explore how a caregiver's window of tolerance directly shapes interactions with a child. And finally I'll consider how we can move away from the rigidity and/or chaos of insecure attachment and into the harmony and coherence of secure relationships.

A SECURE MIND

As a touchstone for our discussion, let's listen to part of a narrative that is very high in coherence. This is a story from my first book, *The Developing Mind*.

"My father was very troubled by his being unemployed. For several years, I think that he was depressed. He wasn't very fun to be around. He'd go out looking for work, and when he didn't find any, he would yell at us. When I was young, I think that it was very upsetting to me. I didn't feel close to him. As I got much older, my mother helped me understand how painful his situation was for him, and for me. I had to deal with my anger with him before we could have the relationship we developed after my teen years. I think that my drive today is in part due to how difficult that period was for all of us."

Like many of us, this woman clearly had a less than ideal childhood, but she can talk objectively about the past, balance positive and negative influences, and reflect on how her understanding has evolved over time. She moves easily from her memories to her reflections on them, and she provides enough detail for me to grasp her experience.

Not every secure narrative is this articulate, but as we turn to the accounts of less securely attached people, you will see that even patients who are highly articulate in their daily lives can become incoherent when they start to tell their life story.

A DISMISSING MIND

You probably recognized some of the AAI questions from my work with Stuart in chapter 6. Let's consider again how he answered my questions about his early life history. Even at age ninety-two, he could readily recall facts about where he lived, the schools he'd attended, major sporting events, and the make and color of his first car. Relationship issues, in contrast, had no place in the story of his life; he insisted he just "did not recall" childhood experiences with his family. What was more, Stuart stated that his family life had no effect on the way he developed—except that his parents had given him "a good education." He seemed eager to move on: "What's the next question?"

How could Stuart know that his family experiences had no impact on him if he could not recall them? This is an example of narrative incoherence—it just doesn't make sense. In other words, Stuart had no evidence to back up his statement. This was especially striking in an attorney, and it revealed a blockage in his narrative integration. The left hemisphere mediates the factual form of explicit memory, which he had in abundance; the right specializes in autobiographical details, which he lacked. Stuart's overdominant left mode had a drive to tell stories, but it was not getting the "goods" from the autobiographical right. As a result, he "confabulated" and made up a story filled with unsubstantiated generalizations such as his childhood being "average" or "fine."

Stuart revealed three characteristics of the AAI category called a "dismissing" state of mind: his lack of recall for details of his relational past, the brevity of his responses, and his insistence that family relationships had no impact on his development. In my own clinical observations, the dismissing state is often associated with left-hemisphere dominance.

Those with dismissing adult attachment were often children with premature autonomy, who acted like "little adults." Reducing participation of the right hemisphere enables them to avoid overwhelming their narrowed window of tolerance for needing others. Leaning to the left is an adaptation, so that they do not feel the pain and longing of missed connections. This is doing the best they can under the circumstances.

What kind of attachment do you think Stuart's son, Randy, had with his dad when he was a child? It's easy to imagine a father who provided well for his son but remained emotionally distant, who might engage him intellectually when he was older but ignored his feelings and had virtually no ability to pick up his nonverbal signals. Having been raised by what his wife, Adrienne, called "the coldest people on the planet," Stuart likely had an avoidant attachment to both parents, and Randy was probably avoidantly attached to him. This is how attachment patterns are passed from generation to generation. Luckily for Randy, he also had Adrienne as a parent, who would have been much more available—physically and emotionally.

The narrative of dismissing adults has a central theme: I am alone and on my own. Autonomy is at the core of their identity. Relationships don't matter, the past doesn't influence the present, they don't need others for anything. Yet of course their needs are still intact—which is why I was able to motivate Stuart to get more connected to his right hemisphere and ultimately to Adrienne. Attachment researchers have monitored stress responses in the skin of adults during the AAI and of infants during the Strange Situation. Even when the dismissing adults discounted relationships in their narrative, and even when the avoidantly attached child ignored the parent's return, the skin test picked up subcortical firings that signaled anxiety.

The child and the adult reveal a similar adaptation: to shut down the attachment system. However, although their cortexes may have adapted with an avoidant and a dismissing stance, their lower limbic and brainstem areas still know that life is about connection. This unacknowledged drive was what propelled Stuart's therapy forward to that moment of meeting when he placed his hand over mine.

For those with dismissing attachment narratives, integration is truly the blossoming of a seed that has lain dormant for decades. The newly developed right hemisphere becomes ready to participate in life and to invite the subcortical connections into the world. It can also link through the corpus callosum to create bilateral integration. Feelings can now become as important as facts. Not everything is recovered: Individuals like Stuart tend not to retrieve autobiographical memories of childhood—these probably were never laid down. But their new narrative integration enables them to create a much richer social, autobiographical, and bodily sense of themselves in the present. "Making sense" goes way beyond having a logical understanding of past events—a coherent story involves all of our senses, head to toe. I could see this happening when Stuart read his journal entries to me, or when he told me that Adrienne's shoulder massage "felt great." It is a wonder to behold, both for the people themselves and for those close to them.

A PREOCCUPIED MIND

Greg would panic when Sara, his live-in girlfriend of four years, came home late from work without calling ahead to let him know her plans. He was a handsome thirty-five-year-old actor, and his sense of uncertainty and anxiety was in stark contrast to his public appearance of self-confidence and success. He frequently questioned Sara's loyalty to him, so much so that she told him she was unable to commit to marriage because of his "insecurities." Greg told me that he too felt hesitant, filled with doubts about whether their relationship would last. On the one hand he knew that Sara loved him; on the other hand he didn't think he could trust what she said. Other women had left him in the past—why would she be any different?

When I conducted the AAI with Greg, I was struck by how this intelligent and otherwise well put together person seemed to unravel at the edges during the process. When I asked him what he remembered about his early relationship with his parents, this is what he replied:

"Well, that's not so simple. I mean, early on I think my relationship with my father was fine. He used to play with me and my older brother on the weekends a lot, and that was good. But when I became older, say as a teenager, my dad couldn't really handle my independence. I kind of lost him, I guess, to his work, I suppose. But my mother was different. She would seem nervous when I'd be with her sometimes, like worried about something I could never understand. It was kind of unnerving in a way that I think made me feel strange. I don't know if she was that way with my brother. I mean she loved us all, but she seemed to favor my brother for some reason. When we would get into fights, even if I lost she'd yell at me. One time I even got hurt and she said it was my fault. Like last week, my mother came to town and visited my brother first even though I live closer to the airport. I mean she still likes him more, and he knows it. When we had dinner last night over at his house, she seemed so proud, so much more, I guess, than she does of me. He has kids, a wife, a house. I have my career, my apartment, a dog, and Sara. Well, you know, it's just not the same."

I had asked Greg for memories of his childhood. But notice how his response slips into the present, so that he's suddenly telling me what happened only a week ago with his mother and brother. This is a different kind of incoherence from the blank spaces in Stuart's narrative. But it also signals an insecure attachment. It is characteristic of the AAI category called "preoccupied," as issues stemming from the past continue to intrude on experiences in the present.

When a child gazes into his parent's face, he is looking for a response that mirrors his own mind. When the communication we receive as a child is open and direct, receptive and attuned, we develop a clear sense of who we are. Our resonance circuits allow us to see ourselves in the face of another and our mindsight lens develops with clarity. But what if that reflective mirror is distorted by the parent's own preoccupations and states of mind? Ambivalent childhood attachment is associated with a history of inconsistent parental attunement combined with episodes of parental intrusiveness. The child cannot see himself clearly in the eyes of his caregiver, and the result is a confused sense of self. A central theme of the preoccupied narrative is: I need others but I can't depend on them.

Another way to understand ambivalent attachment is to talk about "emotional entanglement." A child like Greg is linked to his mother, but he is unable to become differentiated, to have a separate emotional life or identity. The mother's confusing responses, which are driven by her own anxieties, disrupt the balance between differentiation and linkage necessary for integration. Greg becomes filled with his mother's anxiety even when he himself is not feeling anxious. Whatever the internal state he began with, her state has molded his. Instead of two separate individuals who can find a reliable connection with each other, they have become entangled. His mother's inability to see him clearly, her confusions over what we can only imagine might be her own leftover issues, merge into his own mindsight lens. Integration has been blocked, and Greg moves towards the edge of chaos—as in his extreme anxiety when Sara is late. Now Greg can't see Sara as a separate person—one who may have any number of reasons for being late. He can only worry about what her being late says about her feelings for *him*.

The key to growth for Greg lay not in blaming his mother, but rather in understanding the origins of his insecurities so that we could work together to integrate his brain. There is a huge difference between an explanation and an excuse. Greg could grow in his capacity for intimacy by making sense of his life.

My initial goal was to expand the capacity of his middle prefrontal region to monitor, and ultimately to modify, the hyperarousal of his attachment system. (This was the opposite of the shutdown of the attachment system that we've seen in Stuart.) Imagine that Greg is waiting for Sara. As the clock ticks past her expected time of arrival, Greg's leftover theme of emotional abandonment dominates his internal world, and he becomes agitated. At moments of potential loss and uncertainty, Greg's mind bursts through his window of tolerance, pushing him towards chaos. He is now in a complete panic. These leftover issues are "hot buttons," vulnerable points that preoccupy him and dominate his view of his intimate relationships.

To give him some relief from these feelings, I first taught Greg the basic exercise for integration of consciousness using the wheel of awareness. I also taught him how to calm himself by focusing on his breath and visualizing a safe place. Simply learning how to soothe himself was an important step for Greg. Then, resting in the hub of the wheel—which is a metaphor for the prefrontal region—he could distance himself somewhat from his right-hemisphere intrusions of feelings, bodily sensations, and autobiographical images. Rather than becoming swept up into his panicky feelings of insecurity, he could now begin to discern them as just feelings on the rim of the wheel.

I also used the hand model of the brain to help Greg understand how his right hemisphere was flooding his left—and making his prefrontal region unable to cope. Now he could visualize the bilateral integration we were working on. When he learned to "just notice" what his body's sensations were, to honor them without being terrified by them or trying to suppress them, Greg increased his vertical integration. As for his "leftover" issues about his mother favoring his brother, we were able to name the reality of implicit memory—to understand how the deep pain from his past had remained unintegrated by the hippocampus and could be triggered without his awareness, flooding him in the here and now with a sense of being "unlovable." Having named it, Greg could also tame it. That is, with his now-stabilized attention, he could focus directly on these implicit memories and integrate them into more explicit forms.

Greg came to understand that his doubts about Sara were driven by his old feelings of emotional abandonment. These feelings were embedded in his implicit memory and dominated his right hemisphere's data banks. While he did not have "flashbacks" in the

PTSD sense, Greg became aware that these intrusions of intense emotion derived from past events were still driving his life narrative today. And with his newly developed mindsight skills, Greg could begin the crucial process of disentangling his internal concerns from his external reality. His left mode could find a way to sort, select, and sequence his chaotic right-mode data into a more coherent account of his life. Now he could explicitly pinpoint the origins of his worries, which enabled him to approach Sara—and their relationship— in a new way.

Several months into our work together, Greg proudly reported, "Sara told me that she thinks I understand her more—or at least I'm trying to. And she thinks I'm more settled. I think that is good for both of us."

AN UNRESOLVED AND DISORGANIZED MIND

Sometimes childhood relationships leave us with more than leftover issues that preoccupy us and intrude into the present. When our experiences are terrifying and overwhelming, the mind may fragment and become disorganized. The very fabric of our internal world begins to unravel, we become disoriented, and we are unable at times to maintain either a clear connection with others or a coherent sense of ourselves. If the past trauma or loss is not resolved, our internal narrative too will break down. If we try to tell our story to others, we may be overcome by feelings or images that have not found a place in the larger narrative of our lives.

"I lose it whenever he gets upset," Julie told me. She was trying to describe what was wrong in her interactions with her two-year-old son, Pythagoras. A forty-one-year-old high school mathematics teacher, she had come for therapy because she couldn't "work out the equation" of how to raise her first child. She looked older than her age, and her somewhat disheveled appearance pointed at the distress behind her wish for an "equation" that might solve her problem in some neat, organized fashion.

Julie's husband had been married previously, and his two teenage girls were in and out of the house, but they weren't Julie's concern.

"They just don't bother me," she said, "not like Pythagoras. Something about him just drives me crazy." Julie knew it was normal for toddlers to start asserting themselves sometime between their second and third birthdays, but reading about the "terrible twos" hadn't helped her much. Her son's defiant and oppositional behaviors still "pushed all the buttons" in her head and made her "unravel at the seams."

Julie's response seemed more than a parent's concerns about occasionally "going down the low road" and losing her temper. She described a sense of herself "falling apart" when Pythagoras resisted her. She would take him on, getting more and more agitated as they battled over toothbrushing or hair washing. Bedtime had become a nightly crisis, with Pythagoras running around the house and climbing out of his cot until Julie was reduced to tears. It was hard enough being on the front lines after work, she said, but inside she felt like explosions were making it impossible to "think straight." "I feel like I become at first frightened and paralyzed, and then I'm afraid something will snap and I'll scream or yell or, worse, I'll hit him. I feel like I'm losing my mind."

Nothing Julie had told me about Pythagoras made him sound like anything other than a feisty kid with an active temperament. Her husband didn't have the same problems handling him; he told Julie he got a kick out of Pythagoras's spunk and said his son was "all boy." This, of course, left Julie feeling both put upon and alone.

When I conducted the AAI with Julie, she revealed some elements of both a preoccupied and a dismissing stance. If we've had an ambivalent attachment with one caregiver, preoccupation can show up as intrusions of right-mode recollections and emotions, disrupting our left mode's attempt to tell a linear, linguistic, logical, and coherent story. At times Julie's narrative sounded like Greg's. For example, she told me, "My mother was never there for me—at least she couldn't find the time to be only herself with me. I mean she cares about me but she is busy . . . no, more like she's distracted. It's odd." She began to address my question about her past relationship with her mother but she quickly slipped into the present.

The dismissing aspect of Julie's story came out in her lack of recall for many details of her childhood—and in her statement that her childhood hadn't affected her much. Here again is the incoherence that we saw with Stuart: If Julie can't recall the past, how can she be so sure it didn't impact her life?

But something new emerged when I asked Julie the standard interview question about times when she may have felt terrified as a child. First she simply stared at me for a few seconds. Then she said: "Well, I wouldn't say that I had scary experiences so much, as I was frightened, but not that much. There were times, but it happened rarely. And this was when my father, who was an alcoholic, when he'd drink, and I'd be at home, and he'd come in, say, late at night. And most of the time he'd just pass out. But I'd listen closely when the car pulled into the garage, for how loud it was when he slammed the door. If he drank too much, he'd just collapse. And if he didn't drink that much, he'd just be talkative. Somehow I learned to gauge how much he'd had, you know, to work out, well to . . . One time, well, he had that amount just in the middle. And I don't know. He must have been fighting with my mother that night, or something, because she was usually home. But he was mad, and when I saw him in the kitchen he tells me, well, I think . . . He has this knife, this butcher knife. But he's drunk . . . and I think he doesn't mean it, to chase me I mean, to say that I shouldn't be a teenager in the house, not to wear those clothes like that, whatever that means. . . . I ran. . . . I ran into the bathroom, but he broke it down, and I just screamed. . . . I don't remember that night that much. . . . Well, that was scary, I guess, yes, I think."

Julie could hardly get the words out. She was sitting right across from me but I felt I could no longer reach her. She seemed to retreat to a great distance, lost in her mind, filled with the terror of that time. She was no longer recounting her history to me; she was gone, reliving the past in what felt to me like a dissociated state.

It's helpful here to recall the mechanisms of dissociation that I described in chapter 8. During an experience that feels life-threatening, the flood of stress hormones and our internal state of terror and helplessness can shut down the hippocampus. The raw materials of implicit memory are not assembled into more integrated explicit forms. If our awareness becomes divided—as when we focus on some nonoverwhelming aspect of the experience simply in order to survive—we may also encode trauma into this implicit-only form.

Such implicit memories may leave us prone to intrusive feelings, perceptions, behavioral reactions, and bodily sensations. A fight-flight-freeze response from long ago may remain in an implicit-only state, ready for reactivation with minimal provocation under certain circumstances. When these implicit elements are retrieved with a specific trauma-related cue—such as the tears and wails of an upset child—then our own buried distress rises up to flood our here-and-now experience. The trauma-related cues may also be internal: Julie's sense of helplessness in the face of her son's distress, her feeling of impotence to soothe him, may have powerfully evoked the feelings she had when her father came home drunk.

Again, the brain is an associational organ: Neurons that fire together, wire together. Since the brain is also an anticipation machine, ongoing experiences prime the brain to make associational linkages outside our awareness. In Julie's case, her toddler's anger and defiance in response to the inevitable "no" from his mother triggered in her a state of fear bordering on panic. Her reactivity is not experienced as a recollection of any sort. Julie's network of memory associations—implicit here because they were dissociated and unresolved—led automatically to the fragmentation of her otherwise organized brain.

A terrified child is faced with a biological paradox. Her survival circuits are screaming, "Get away from the source of terror, you are in danger!" But her attachment circuits are crying out, "Go towards your attachment figure for safety and soothing!" When the same person is simultaneously activating the brain's "go away" and "go towards" messages, this is fear without solution—an unsolvable situation. Here the self of the child is not *disconnected*, as in avoidance, or *confused*, as in ambivalent attachment. Instead the child's sense of self becomes *fragmented*. This is called disorganized attachment. It is characterized by unresolved states of trauma and loss, and by the kind of dissociation that Julie seemed to be experiencing.

So let's summarize. The presence of unresolved trauma or grief in the mind makes for a narrative prone to disorientation and disorganization during those specific moments in the story that are related to the terror or the loss. Researchers call this an unresolved/ disoriented pattern. Its theme is something like "At times I fall apart, so I can't depend on myself."

With unresolved trauma, an otherwise coherent story can suddenly become fragmented as the speaker breaks through the window of tolerance. This is a sign of disintegration. Similarly, a parent's relationship with a child may ordinarily be attuned and secure. But when certain specific stressors arise, the holes in his or her ability to cope are revealed, the window of tolerance narrows dramatically, and behavior falls apart. Under these conditions unresolved states can create a low-road reaction, and we fly off the handle or lose it. Julie had good reason to fear that she might hit Pythagoras or terrify him by screaming at him. When such low-road eruptions are intense or frequent, they can be traumatizing for a child. And unless the resulting disconnections are repaired, the child may develop a disorganized attachment that mirrors the parent's own childhood experience.

I began a process of slowly helping Julie examine her experiences with her father. When we started, she had no coherent narrative that could help to distance her from her implicitly encoded memories. There was no context that could help her understand her reactions as remnants of a terrifying past. Instead they were the terrifying realities of her here-and-now relationship with her son. She was stuck on the rim of her mind's wheel of awareness, completely out of contact with the mindful presence of the hub.

As Julie and I explored the connections between her past and present, certain narrative themes began to emerge. She saw that feelings of being out of control with Pythagoras actually came from her experiences with her father. Feelings of betrayal also emerged in her therapy, betrayal not only by her father, but also by her mother, who had turned a blind eye to the abuse Julie suffered during her father's drunken rages. She had many reasons not to remember the details of those times, which no doubt contributed to the dismissing elements of her narrative. No wonder she had taken refuge in her left brain and in the abstract world of mathematics. But now she was beginning to see the logic and history behind her previously irrational and inexplicable reactions to her son.

During this period, Julie joined a group of mothers with toddlers and found their shared perspective—with its mixture of exasperation and humor—very helpful. She also attended Al-Anon meetings, which helped her to understand and share the experiences she had had with an alcoholic father. But Julie seemed to benefit even more from the internal work of mindfulness practice and journal writing. Writing in a journal activates the narrator function of our minds. Studies have suggested that simply writing down our account of a challenging experience can lower physiological reactivity and increase our sense of well-being, even if we never show what we've written to anyone else.

One day Julie came to a session and told me that her son had had a tantrum. Then she said, "I could see my mind getting ready to blow, seeing Pythagoras's furious face becoming, literally, my father's. I knew then that I was in trouble—I was seeing double." After writing about this encounter in her daily journal entries, and choosing to reflect on it during her mindfulness practice, she began to see such challenging moments as opportunities. Several weeks later, she commented, "I know this sounds crazy—but I'm now grateful to Pythagoras for being as strong as he is. I have to take care of my own issues, heal myself, and not fall into that pit of seeing this as his problem and not mine. I know there is work to do, but I think I know the way to begin now."

Therapy widened Julie's window of tolerance so that she could bring her terrifying right-hemisphere images into relationship with her left brain's ability to understand them. Therapy also gave her an external source of safety, a protected space and a personal connection with someone—her therapist, me—whose goal was to help her see her mind without the distortions of the past. This was essential at first, although she gradually learned that her support group, her friends, and her husband could be sources of sustenance as well. And as the hub of her mind strengthened, she could draw the raw data of her experience from the rim and assemble them into the coherent story of who she was and who she wished to be.

Having the courage to approach and not avoid her past trauma allowed her to become free of its implicit grip on her mind. Julie harnessed all the other domains of integration we have discussed so far—consciousness, vertical, horizontal, and memory—in order to achieve narrative integration. She could now become truly present in her own life, and it was a fine thing to watch Julie gradually gain confidence in herself as a mother. She was learning that she could, indeed, depend on herself.

It is not only Julie who will benefit from this healing process. Pythagoras will be able to form a secure attachment with her, and this will be a source of resilience for him throughout his life. Julie will have stopped the passage of abuse and terror from one generation to the next once and for all. This is why mindsight is important not just for our own well-being, but also for what it enables us to give to our children (and others). It is never too late to heal the mind and to bring to ourselves and to those around us the compassion and kindness that arise from that healing and integration.

LIGHTING UP OUR LIVES

When we see the mind of another person we bring the qualities of being present—curiosity, openness, and acceptance—into our relationships. These qualities seem to me to be the essence of that overused, often misunderstood word: *love*. I propose that this stance of curiosity, openness, acceptance, and love is at the heart of secure attachments. And this stance is the felt sense you pick up from a coherent narrator's relationship with himself.

Self-compassion and self-acceptance emerge quite seamlessly from the "secure attachments" that are the result of consistent, continuous and caring connections with our caregivers in early life. But they can also emerge from an "earned secure attachment," as with my friend Rebecca, who had such a difficult childhood. When, as Rebecca says, we feel that we are "inside the heart" of another, the candlelight of love glows within and illuminates our lives.

For most of us it is our parents who light that candle. For Rebecca it was her aunt. Feeling felt by her aunt gave her a sense of herself as real and valuable even in the face of the chaotic parental environment at home and ultimately allowed her to arrive at a coherent narrative of her life story. If we have a positive relationship with a relative, teacher, counselor, or friend, the path is set for us to create a positive relationship with ourselves. We can then light our own interior world with mindsight and see our lives as stories that hang together and have meaning. This is one reason I always try to encourage teachers and therapists to offer a solid, attuned connection to their students and patients. Wonderful things happen when people feel felt, when they sense that their minds are held within another's mind.

My friend Rebecca now has children of her own—children who have the gift of a mother with whom they can have an open, loving relationship. If you saw Rebecca with them, you'd never guess what a painful childhood she had. Early experience is not fate: If we can make sense of our past—if we integrate our narratives—we can free ourselves from what might otherwise be a cross-generational legacy of pain and insecure attachment. Rebecca has always served as an inspiration for me for how taking responsibility for one's own mind can lead to liberation of the self, and to the ability to love and nurture the next generation.

10
Our Multiple Selves
Getting in Touch with the Core

MATTHEW KNEW HE WAS IN TROUBLE when, without warning, his fourth girlfriend in five years slammed the door of his house behind her, thus slamming it on their relationship as well. At least this was the way he initially told his story. I'd soon discover that within Matthew's mind there had been many warning signs that something was awry, a feeling that he was watching the reenactment of a pattern of behavior that was out of his control.

Matthew was a forty-year-old investment banker renowned in his field for his affable nature and for his shrewd, profitable business decisions. His public persona was that of a confident, easy-to-get-along-with guy, yet in his private life he didn't seem to be able to maintain the intimacy that he said he wanted. Something would "take him over" and he would create his own worst nightmare, lose his partner, and be alone, yet again.

At work, Matthew made decisions concerning large sums of money without hesitation or second-guessing. He was sure of himself and clear in his thinking. But when he first came to see me, this financial success was like a surface film over a deeper pool of pain of which no one, not even Matthew himself, seemed to be fully aware. He certainly had no idea why there was such a split in his experience: solid as a rock at work, a broken branch at home.

Matthew had made many people wealthy—his reputation was founded on results, which led to still more investment opportunities. It was a lucrative, upward spiral that yielded both financial rewards and the social benefits that made Matthew what he termed a "hot commodity" in the city's singles scene. The financial markets were flying high during those years, and Matthew's women were like the high-profile, expensive properties he traded at work—glamorous, highly desirable, and accessible only to a few. But this business strategy yielded strangely barren results outside the office.

Matthew's large numbers of dates, like the abundance of his clients and money at work, had in fact done little to make him feel that he was worth all the attention. After a month of therapy he acknowledged to me that despite his professional success, inside he "felt like an impostor just waiting for them to find me out."

When Matthew was younger, he'd longed for "the chase." Seeking out women, having sex, and then never seeing them again—he was the ultimate "one-night-stand kind of guy," he told me. After his late twenties, he said, this routine bored him and he came to realize "how empty these sexual encounters were. I had conquered, but I had nothing. It was terribly painful." At thirty Matthew decided to make a change. But things weren't working out as he had hoped.

In the years following his thirtieth birthday, Matthew's revolving door of girlfriends added up to a dizzying list of failed attempts to achieve something he said he wanted but could not really define. Matthew sought these women with tremendous energy, and he was glad to have gotten past the one-night encounters of his younger days. But after he'd won them over and they came to know and accept him, something changed inside him. Instead of the deepening attraction he'd hoped he would feel for a woman after the first months of dating, he would feel more and more repelled by her caring.

During the initial period with a new woman, Matthew would find himself electrified in her presence. He'd send flowers and small notes, make surprise visits to her workplace or home. His infatuation had an exhilarating drive that he found "addictive." Matthew loved a challenge and was drawn to mastering the near impossible. This was his passion at work, and this is how he carried out his romantic life. He'd pick a woman whose public profile and good looks were something that, even given his own status, would seem to make her "out of reach." This disparity of worth—in his own mind, of course—galvanized his resolve to make her interested in him. Occasionally he sensed that he was becoming lost in some familiar chase, less interested in the person than in the pursuit. But he was driven by some potent elixir that had little to do with intimacy and everything to do with the challenge.

LOST IN FAMILIAR PLACES

One of my first hypotheses was that Matthew was primarily—as he said—addicted to the thrill of the chase. In brain terms, this would involve surges in the release of the chemical messenger dopamine, which plays a central role in drive and reward. All addictive behavior, from gambling to the use of drugs such as cocaine and alcohol, involves activation of the dopamine system. Rats will consume cocaine in lieu of food or water. Dopamine activation is so rapid and intense with cocaine that no other activities or substances can compete with it. The neural areas initiating reward seem to overwhelm the prefrontal areas that regulate our more complex behaviors, so that instead of our being able to choose our actions, the drug chooses them for us. The reward circuits take over, and our conscious cortical mind becomes a slave to the addictive drive.

But I soon realized that Matthew's surge addiction was only one element of his relationship history. While a simple drive for dopamine might have led to the kind of impulsive and promiscuous sexual encounters without regard to safety or selectivity that were typical of Matthew's behavior in his twenties, his pattern had changed since then. Now he was strategizing and carrying out long-term plans for conquest. He could wait, plan, and pursue his romantic interests with great patience—not a dopamine-surge-driven kind of behavior pattern. As he and I explored his more recent romantic history further, Matthew himself stated that he felt that proving his ability to win over "high-profile women" would somehow establish his worth. Perhaps this goal—of using the people he associated with in order to convince himself, and others, of his value—is not so uncommon. But the specific source of pain for Matthew was that none of these relationships ever lasted. Matthew could not seem to get what he wanted. And no matter how hard he tried, he couldn't even get what he needed. In Matthew's life so far, the Rolling Stones were wrong.

Often the women Matthew chose would initially be cool and indifferent, but soon some would become more affectionate and begin to care about him deeply. Then, rather than seeing this as a sign of a successful relationship—or even a sign of his worth—Matthew would lose his drive to be with the woman and suddenly begin to act in ways that would cause her to leave him. Once a new girlfriend showed signs of liking him, Matthew's sexual attraction to her started to wane. Even more, if she showed affection for him outside the bedroom, he began to feel repulsed, even nauseated, by her caring behavior. If he tried to go through the motions in bed, his lack of sexual arousal became all too evident. He became self-conscious, and intercourse was sometimes not possible at all. Then Matthew would find himself doing things to pull away or to alienate her. He'd let loose with irritation and frustration. If she responded with concern, he'd up the amplitude of his disgust. He'd find himself not returning her phone calls or ignoring her when they were together. Usually at that point things just fell apart.

A theme began to emerge: Matthew was stuck in a maddening cycle of contradiction, repeatedly sabotaging the very thing he believed he was trying to achieve. My clinical impressions had come together as well. Matthew seemed to be trying to undo some profound sense of inadequacy. When you do not value yourself, the positive appraisal of others can, ironically, become a painful source of discomfort. As Groucho Marx famously said, "I don't care to belong to a club that accepts people like me as members." Woody Allen, who quotes Groucho in his classic film *Annie Hall,* might have put his arm around Matthew and told him to lighten up, but the pain of rejection was no laughing matter for Matthew. He frequently found himself alone, rejected by the very people he had expended huge amounts of effort, time, and money to pursue. Once they had invited him in, he would make his exit.

AN UNSOLVABLE CONFLICT

Matthew's answers to the Adult Attachment Interview opened the door to his inner world. His father had chronic lung problems, emphysema and asthma, and was bedridden for most of Matthew's childhood. Matthew could recall being brushed aside, told by his mother that he shouldn't bother his father, that if he did anything to upset his father it would "kill him." His two older sisters were busy with school and babysitting jobs. His mother, a talented pianist before she married, had taken a job as a middle school music teacher when his father could no longer work. She made no secret of her frustration and anger at her situation, and as Matthew came to see in retrospect, she was also profoundly frightened and alone.

Early in our discussions Matthew described mainly a feeling of distance from his mother. But one day we entered deeper waters. We were trying to explore why he so often felt anxious and irritable when he was out to dinner with the current woman in his life. His eyes filled with tears, and he began to sob. At some point, he told me, his mother had become convinced that his father's illness was caused by poor nutrition. To "help keep us all healthy," he said, she'd prepare mounds of food for the family, food his father didn't have the energy to eat. When Matthew couldn't finish what she'd piled on his plate, she'd banish him to his room. Later, when his sisters left for their babysitting jobs and his father was asleep, she'd come to his room and berate him for his failings. And sometimes she would use a belt on him "to let me know how much she cared for me."

During these early sessions when we explored his past, Matthew would at times enter a shutdown mode, a form of "collapse," as he later called it, in which he felt "stuck and unable to move." He'd stop speaking and just stare out into the room, seemingly lost in thought. When he came out of it and tried to tell me about that internal para-lyzed state, what he described sounded like the "freeze" arm of the fight-flight-freeze response. It was as if his brain had assessed a life-threatening danger, where collapse and helplessness were the only possible response.

But easy-to-get-along-with Matthew was also starting to show me the "fight" in his response system. He'd tumble down the low road in response to some minor irritation. Once I forgot to put my cell-phone on "vibrate" before our session and the ringtone angered him. "I'm paying for this time, and I want to know that you respect it," he snapped. His reaction to being interrupted was understandable, but his hostility was, as he later admitted, "through the roof."

Matthew's mother had created the biological paradox of disorganized attachment in her son: He was scared of her and he was driven to escape the source of fear. At the same time the attachment circuits in his brain were driving him towards his attachment figure for soothing. As I discussed in chapter 9, the problem is that those two states embed two opposite drives aimed at the same person at the same moment. Such a conflict is not resolvable—it is the "fear without solution" that Julie faced with her alcoholic father, the fear that is at the heart of a disorganized mind.

These repeated episodes with his mother during his preadolescent years not only were terrifying at the time, but they also seared yet another state into Matthew's brain: the state of shame.

SHAME ON THE BRAIN

Imagine a car with the accelerator smoothly functioning. When we need to be seen and understood by others, our attachment circuits are revved up; we are in a state of seeking connection. And when our need is met, we move forward happily through our lives. But if we are not seen, if our caregivers do not attune to us, and we are met with the experience of feeling invisible or misunderstood, our ner-vous system responds with a sudden activation of the brake portion of its regulatory circuits. Slamming on the brakes creates a distinctive physiological response: heaviness in the chest, nausea in the belly, and downcast or turned-away eyes. We literally shrink into ourselves from a pain that is often beneath our awareness. This nauseating and jolting shift occurs whenever we are ignored or given confusing signals by others and it is experienced as a state of shame.

Shame states are common in children whose parents are repeatedly unavailable or who habitually fail to attune to them. When shame from nonattuned communication is combined with parental hostility, toxic humiliation ensues. These isolated states of being—shame intensified by humiliation—burn themselves into our synaptic connections. Now the slammed-on brakes of the freeze response are painfully combined with the floored accelerator of rage. In the future, we'll be vulnerable to reactivating the state of shame or humiliation in contexts that resemble the original situation—as happened when Matthew needed to be seen and cared for by a female, whether it was his mother when he was a child or his girlfriends as an adult.

As the child grows older and the cortex develops more fully, the state of shame becomes associated with a cortically constructed belief that the self is defective. From the point of view of survival, "I am bad" is a safer perspective than "My parents are unreliable and may abandon me at any time." It's better for the child to feel defective than to realize that his attachment figures are dangerous, undependable, or untrustworthy. The mental mechanism of shame at least preserves for him the illusion of safety and security that is at the core of his sanity.

It is here that we can begin to see the developmental and neural origins of many of Matthew's underlying issues and of his states of humiliation and rage, fear and anxiety, shame and frozen terror: fight, flight, and freeze. Because he hadn't integrated these reactive states into his own narrative, he was as helpless in dealing with them as he had been when he was a little boy and his mother entered his room with a grim face and a belt.

In life we do the best we can, but the shame-based conviction that we are defective, which often goes underground, beneath our cortical consciousness, can sabotage us if it remains unconscious. Although such subterranean shame can compel us to succeed—to prove we are good and worthy of others' respect and admiration—our developmentally ancient feelings of being damaged goods are likely to surface at any hint of stress or failure, and we may become highly reactive in order to keep others at a distance. We need to prevent them—and ourselves—from becoming aware of our shadowy past, the hidden truth of our rotten self. In our personal lives, intimacy is compromised because the closer others come to the real self beneath our public persona, the more vulnerable we feel and the more alarmed we are that this secret truth about our defective nature may be revealed.

Such a profile can help explain why Matthew worked so hard to win over women who initially found him uninteresting, the "impossible ones" who exercised such a powerful draw on his attention because they reminded him implicitly of his mother. Matthew engaged in this conquest-acceptance-repulsion cycle as if his life depended on it. In some ways, his life as a child did depend on finding a way to convince his mother and his father that he was worthy of their love and attention. This drive to prove, to convince the nearly inconvincible, had remained his familiar place even in adulthood. He'd find the challenging symbol of his mother—his difficult-to-get girlfriends—and woo them to prove his worth, striving to assuage a feeling of shame of which he was not even aware.

But as soon as a woman became affectionate, the game-to-prove was won and the real danger began. There was nowhere to hide, nothing to do but run—or get the woman to leave. Painfully, the isolation he felt as a child was being disastrously re-created as an adult. This is how shame repeatedly led to Matthew's becoming lost in those familiar places. He was caught up in the cycle of isolation, and his alternating states of attraction and repulsion had driven him to a dead end.

MULTIPLE SELVES

In earlier chapters we've seen that dissociation has a broad spectrum, ranging from everyday absorption in a daydream to a psychiatric disorder. In disorders of dissociation, the normal continuity across consciousness is disrupted. When memory is fragmented, patients lose their sense of a coherent self, they lose a sense of being connected to their body, and they feel unreal. At the extreme end of the spectrum of dissociation is a condition called "dissociative identity disorder," also known by its former name, multiple personality disorder.

Although Matthew's state shifts created a feeling of being "taken over" by something beyond his control, he did not feel as though he disappeared, lost memory, or lost touch with reality—as happens in dissociative identity disorder. He did not experience these states as different from "him." In fact these states of mind had for a long time simply felt like parts of his personality, "natural" responses to whatever was going on.

As we worked together and I heard more details of his relationships with women, he revealed quite dramatic states of mind—such as rage, shame, and fear—that were well-established, frequently repeating patterns, but also quite unintegrated in his life. When I say his states were unintegrated, I mean that they triggered automatic and unwanted behaviors that did not respond to his conscious efforts to change them, and they created serious dysfunction in his social life and distress in his internal world. In short, when a person's states are unintegrated he is internally distressed and inclined towards chaos or rigidity—or both; he is behaviorally compromised as well, unable to be flexible and adaptive in his interactions with others. The kinds of abrupt shifts from one intense emotional state to another that Matthew was experiencing are characteristic of unresolved post-traumatic adaptation.

Another way to understand Matthew's situation is to look at it from the perspective of normal development. Early adolescence is filled with tensions among states, a conflict that is initially out of awareness. By mid-adolescence, these conflicts become more conscious, but teenagers still lack effective strategies for resolution. An adolescent may act one way with friends, another way with siblings, teachers, parents, or members of his hockey team. Clothes, hairstyle, and manners become symbols of different roles, and of the intense conflicts among them. By late adolescence, most young people develop more effective ways to deal with these unavoidable tensions across states. Healthy development is not about creating a single "self" that is a homogenized, uniform entity. Rather, healthy development involves coming to acknowledge, accept, and then to integrate one's various states: to discover how disparate states can link, and even collaborate as a unified whole composed of many parts.

Matthew, however, had not mastered this essential dimension of his development. Many research studies suggest that when such collaboration across states does not occur, adolescents develop mental dysfunction, such as anxiety, depression, or identity issues. On the other hand, adolescents who learn to negotiate across the different states, and who find the settings, friends, and activities where their multiple selves can feel at home, continue to develop and thrive. Once again integration comes along with well-being.

STATES OF MIND

By now you may be asking, what exactly are these many "states" or "selves" that each of us have? In brain terms, a state is composed of a cluster of neural firing patterns that embed within them certain behaviors, a feeling tone, and access to particular memories. A state of mind makes the brain work more efficiently, tying together relevant (and sometimes widely separated) functions with a "neural glue" that links them in the moment. If you play tennis, for example, each time you put on your shorts and shoes, pick up your racket, and head for the court, your brain is actively creating a "tennis-playing state of mind." In this state you are primed to access your motor skills, your competitive strategies, and even your memories of prior games. If you are playing a familiar opponent, you'll recall her moves, her strongest hits, and her weak spots. All of these memories, skills, and even feelings—of competition and aggression—are activated together.

Sometimes the adhesive holding a state together is flexible, enabling us to be receptive and open to bringing in new sensory data and new ways of behaving. You can learn from your opponent and respond to her game as it unfolds. Your state of mind is unique to this moment in time, a one-of-a-kind combination of neural firings, yet it is influenced by the past. You are ready and receptive.

But some engrained states are more "sticky" and restrictive, locking us into old patterns of neural firing, tying us to previously learned information, priming us to react in rigid ways. This locked-down state is "reactive"—meaning that our behavior is determined in large part by prior learning and is often survival-based and automatic. We react reflexively rather than responding openly. An experienced tennis player who feels threatened by the skills of a younger opponent may lose focus if she takes the lead, and if he fails to adjust his play he may lose the game he was sure he would win.

With any activity, we can be receptive or we can be reactive. These qualities of receptivity or reactivity can appear in any state, whether it's helping a child with homework, giving a speech, shopping for clothes, or making love. Each of these activities, if repeated, pull together feelings, skills, memories, behaviors, and beliefs into a cohesive whole. Some states are engaged frequently enough to help define the individual; these so-called self-states combine to create our personality. These are the many selves, whether receptive or reactive, that make up the person we call "myself."

The self-states that were activated when Matthew was with a woman were organized around shame and around his traumatic fight-flight-freeze survival reactions. They primed him to respond in characteristic ways, but in this situation he was functioning on automatic pilot, driven primarily by old implicit learning. When a woman became affectionate and he found himself pulling away, he'd have no awareness of the states that had taken over his mind.

I want to be clear, however, that self-states are part of everyone's life, even if we have no history of trauma. Matthew would often arrive at a therapy session in what we might call his business state. He'd be energized and excited by a successful deal, glowing with confidence, and happy to share his success with me. But as soon as we turned our attention to his latest relationship, his enthusiasm and confidence dropped away and he entered a state of anxiety and uncertainty. This was painful but normal—as any therapy patient will recognize.

Many self-states are organized around our basic biological drives, sometimes called "motivational drives," which originate from our subcortical circuits and are shaped by our regulatory prefrontal cortex. A list of such basic drives includes exploration, mastery, play, reproduction, resource allocation, executive control, sexuality, and affiliation.

If I love cricket, for example, my motivation to join the department's team after work is multilayered: It satisfies my basic drives for affiliation and play. Every time at bat, every adjustment on the field, engages my drives for executive control and mastery. The uncertainty and openness of the game meet my need for exploration. Then when the game is over and I'm tossing the ball around for fun, my resource allocation circuits may remind me I'm hungry and need to get some rest before the workday tomorrow. I head home to eat and sleep after a full day.

These motivational drives obviously pull together input from the body, the brainstem, and the limbic areas, but the cortex also plays an important role in self-states. What the cortex does is easiest to understand if we take another look at its basic anatomy.

TOP-DOWN AND BOTTOM-UP

Six cells deep. That's it. Our powerful perceiving-and-planning cortex is organized by stacking six neurons on top of one another and clustering these piles—or "cortical columns"—like an interconnected honeycomb. Cortical columns close to one another coordinate the information flow in a similar modality: Vision, for example, is carried out by columns located at the back of the cortex, in the occipital lobe; hearing by columns on each side, in the temporal lobe; touch by columns higher on the side, in the parietal lobe. When we plan a motor action, columns in our frontal lobe become activated. And when we form an image of our own mind or the mind of others, columns of neurons in the middle prefrontal area are firing.

To understand how states of mind are shaped by prior learning, we need to grasp another amazing fact: The flow of information through a cortical column is not just from input to output; it is not one-directional. Cortical column flow is bidirectional. This is one important key to Matthew's states of mind—and states of mind in general.

Incoming sensory data rises through the brainstem, enters the cortex at the bottom layer of neurons, and makes its way upward. This is called "bottom-up" information flow. When a toddler comes face-to-face with a rose, he may be attracted first by its bright red color, then sniff its fragrance (smells are routed directly from the nose to the cortex), touch its petals, and even try to eat one (until his mother notices). This is as close to direct perception, or pure bottom-up experience, as any of us is likely to get.

But if we've seen a rose before—and for many adults, this applies if we've seen *any* flower—a rich store of memories from similar experiences is activated by the rose. Prior learning sends related information down from the top layers of our six-neuron-deep column to shape our perception of what we are seeing or hearing or touching or smelling or tasting. There is no "immaculate perception"; perception is virtually always a blend of what we are sensing now and what we've learned previously.

See if you can visualize this: Sensation moves upward from neuron layer 6 to 5 to 4. These "bottom-up" inputs meet the "top-down" influences coming down from layer 1 to 2 to 3. The top-down influences include our present state of mind, our memories, our emotions, and our external setting. In the middle, at neurons 3 and 4, the two information streams mingle or crash. What we become aware of is not what we sense but what emerges from this confluence.

Suppose, for example, that you see me raise my hand over my head. If you and I are on a New York City street, you'll probably assume that I'm hailing a cab. If, on the other hand, we're in a classroom, you'll know that I want to ask a question or make a comment. Same gesture, different contextual setting, different prior learning. You wouldn't have to think about the meaning—you would just automatically "know" what my moving hand meant. This is the benefit of a state of mind as it creates an efficient top-down filter through which we interpret the world. (This is also an example of our mirror neurons at work, using prior learning to determine the intention of an action.)

But states of mind can distort our perceptions, too. If you had been physically abused as a child, and the setting was more ambiguous—say we're at a party and are having a heated discussion—interpretation would become more difficult. In that context, if I raised my hand quickly to emphasize a point, you might fear I was going to hit you. Your top-down cortical flow would dominate your bottom-up visual input and you'd completely misperceive my intention. Here your mirror neurons would distort your ability to see me clearly. This is how leftover issues and unresolved trauma can create a reactive top-down filter. Otherwise you might either enjoy our hot argument and be receptive to my ideas, or just decide to walk away. Again, same gesture, different outcome.

Learning how perception is shaped by the architecture of the cortical columns helped Matthew begin to make sense of his unintegrated states of mind. He followed closely when I described how interactions with our parents shape our neural development and top-down filters, and he was intrigued by the notion that it's normal to have different—and even conflicting—states of mind. The challenge, he came to understand, is not to get rid of top-down influences (we can't), but to become aware when a given self-state is reactive from the past and not receptive to the present.

I also wanted to make sure he understood how strongly top-down flow can dominate bottom-up input. When we are on autopilot, our awareness "believes" what it perceives. There is no mindsight, and our state-dependent perceptions, emotional reactions, beliefs, and behavioral responses are felt to be justified, equated with absolute reality, not discerned as just activities of the mind. Before therapy, Matthew's "intuition" and "gut feelings" were telling him his girl-friends were repulsive and he was totally convinced by this distorted top-down communication. Top-down forces can shape what we think in the blink of an eye, distort the reliability of our instinctual responses, and challenge our most cherished sense of free will.

So what is left for us to rely upon? How can we know who we really are, what is good for us, what is true? If we have so many states of mind, which one defines us, and which should we choose to be? The answers to these fundamental questions emerge from state integration.

STATE INTEGRATION: INTER, INTRA, AND WE

State integration involves linkage in at least three different dimensions of our lives. The first level of integration is between our different states—the "inter-" dimension. We must accept our multiplicity, the fact that we can show up quite differently in our athletic, intellectual, sexual, spiritual—or many other—states. A heterogeneous collection of states is completely normal in us humans. The key to well-being is collaboration across states, not some rigidly homogeneous unity. The notion that we can have a single, totally consistent way of being is both idealistic and unhealthy.

The second level of state integration takes place within ("intra-") the given state itself. A state needs internal coherence in order to function—to achieve its goals effectively and without internal disintegration. For example, I decide to join a gym to increase my physical fitness. If I've never allowed myself to be athletic, if as a kid I was mocked for my clumsiness and still feel that old fear and confusion, then I'll need to do some reflective work with myself. Otherwise, that leftover baggage is likely to sabotage my goal. I probably won't enjoy what I'm learning, and I'll find myself going to the gym less and less frequently.

The third dimension of state integration involves who we are in relationships. Our history shapes how our sense of being an "I" can become a part of a "we-state" without being obliterated by this join-ing. Becoming open to this we-state of mind requires us to be vul-nerable and receptive—qualities that are challenging for many of us. There was no safe we-state available to Matthew as a child, and he was finding it impossible to achieve one now.

Matthew and I had our work cut out for us in all three of these dimensions.

UNTANGLING THE KNOT OF SHAME

You might ask, "Well, why doesn't Matthew just get rid of those shame states?" The results-oriented businessman in Matthew had a similar impulse; he wanted to "wipe out" the aspects of himself that he could not bear. Unfortunately, that divide-and-destroy approach simply does not work. Each of our states is fulfilling some kind of unmet need. In order to begin inter-state integration, it is important to approach these deeper needs, identify them, and find more adap-tive and healthy ways to meet and satisfy them.

What if our basic motivational states are in conflict? Some states collaborate well (sexuality and play, for example) but others clash. So we need to find a way to embrace strong motivational drives that exist side by side: our need for focused mastery and our need for open-ended play; our drive to monitor our resources (of time, energy, money, food, etc.) and our drive to reproduce (children cost us a lot in energy, money, and food—which is as true for modern city dwellers as it was for early humans); our need for exploration (to fol-low our individual creative interests) and our need for social affilia-tion (to remain members in good standing of our family or community requires fitting in with others). These built-in contradic-tions are one reason that balance and variety are so essential to our health.

Here is how Matthew and I approached the conflicting states that were tearing him apart.

Matthew and I could easily identify the part of him that desperately wanted to have a partner in his life. "I am so done with my twenties," he said, looking back at his sex life after high school. "I'm looking to settle down now, but I just can't seem to find the right person."

In fact, at this stage of his life, the "right" person Matthew needed to find was himself. While one self-state wanted closeness and intimacy, another self-state needed to protect his vulnerability, and yet another needed to prove his self-worth. These states were clustered in different firing patterns in Matthew's brain—filtering his perceptions quite differently from the "closeness-seeking" state.

Think of Matthew's cortical columns: In a closeness state, he sees this attractive woman, his girlfriend. In that state, he perceives her as "right" for him in so many ways—her intellect, her sexuality, her personality, her humor. These are the reasons he is drawn to her. But then as she grows more fond of him, as she comes to like the wonderful person that he is (and he can be quite wonderful and kind), "something" shifts in him. That shift, Matthew and I came to understand, was the activation of another set of self-states.

Shame organized Matthew's self-states into a number of related but distinct clusters. One was simply protective—if his girlfriend is interested in him she might get to know him and discover that he is, deep inside, a jerk. Better to leave her before she finds out. This state also protected him from the threat of sexual failure: If he really wanted to be close to a woman, if their relationship truly mattered, then the idea that he might "screw things up" was so distressing that it was better to end the relationship before he did—just as, in his twenties, it had been better to have sex with women he didn't care about so that it didn't matter how he did. This was one of the many-layered reasons Matthew became self-conscious and lost interest sexually once his girlfriend liked him "too much."

Another shame-based self-state was more punitive. If she really liked him, how could he ever forgive her? Does that line of reasoning sound irrational? Here's the logic: If a woman likes me, there must be something wrong with her. So why would I want to be with her? Shame explains that equation. When we have a deep belief that the self is defective, all of these "irrational" responses make total sense.

Matthew's conquest drive was yet another shame-based self-state. When he chose "hard-to-get" women, a part of him felt totally compelled to win them over. He had never chosen women who liked him up front, never. Even women who were just neutral didn't attract his interest. For a self-state still trying to conquer an old trauma, the "best" approach was to re-create neural firing patterns as close to the original as possible. One clinical term for this is *traumatic reenactment*. In brain terms, he was seeking memory triggers to activate self-states that were always ready to engage with distant and potentially abusive, mother-resembling women. Matthew had a knack for finding them—at least on the surface.

But another self-state from Matthew's childhood was still active as well: a young state simply needing love and connection. A number of his girlfriends had glimpsed this state, and it would open their hearts. It was beautiful to hear about these precious moments when Matthew could accept a woman's affection, even though his shame-based self-sabotage would soon return.

So what could we do? State integration required that Matthew stabilize his mindsight lens just as Jonathon did in chapter 5. We took a break from romance, and over the next few weeks our sessions focused on teaching him the various reflective techniques. Matthew liked the wheel-of-awareness metaphor and the idea that mental exercise could strengthen the hub of his mind. Although he was skeptical at first, he also found the body scan to be helpful. The intensity that he brought to these practices felt like the state he had at work: We could name a goal and he would pursue it with highly focused energy.

But Matthew soon found that accepting whatever arose required a new kind of awareness. It was difficult for him to become open to his inner world without trying to control it. Take the strong feeling of repulsion that would arise when he was with a woman: He needed to observe its emergence; he needed to be objective about this being just a part of who he was; and he needed to remain open to the deeper pain that was driving it.

When I introduced Matthew to the "stay with that" practice (the one I taught to Anne in chapter 7), he was intrigued by the seeming paradox of using the strength of his mind to just be curious and open, to be accepting of his inner world. Curiosity, openness, and acceptance, I told him, were in many ways the fundamental ingredients of love. Those, Matthew said, were the very things missing from his life as a child.

As with Anne in chapter 7, therapy for Matthew required a dual focus of attention—one tracking his moment-to-moment experience in the past, or with the women he was now having relationships with, the other grounded firmly in the present, with me in the room with him. Through many challenging sessions, we came to see how raw his childhood experience of rejection and terror still was, and how much support he needed to "stay with it." Matthew also needed many of the skills of integration—bilateral, vertical, memory and narrative—to move these raw implicit recollections into their more flexible explicit forms.

In one session, Matthew recalled going into his father's room to see if he could play with him. He must have been about six years old. His mother rushed in, grabbed his arm, and pulled him out the door. "How many times do I have to tell you not to bother him with your nonsense?" she demanded. Now, in my office, his arms began to quiver; he saw his mother's face and recalled how terrified he had been by her fury. I asked him to "stay with that fear," holding it in the front of his mind. We sat together with his terror and experienced how it unfolded into sadness. Matthew began to cry.

I showed him a way to hold himself, one hand over his heart, the other over his abdomen. For many people, this provides a powerful experience of self-soothing. Matthew had never had a way to take care of the pain of his shame without trying to escape from it. I hoped this technique would help him widen his window of tolerance. After a few minutes, he said it helped a lot, and we discovered that for him—as it is for me—the left hand over the heart is the most soothing. (It's actually the other way for most people.) Embracing himself, he could also embrace his implicit memory of a child who longed to be loved, to be accepted, to be seen for who he was.

Once Matthew was calmer, more memories came. He told me how he had taken on a newspaper route as early as he could, at age twelve. He'd used his first earnings to buy his mother a blender so she could make milk shakes for his father. "She hardly said thank you," he said. "I did well in school, I'd buy her flowers, later I spent the weekends washing cars so I could give her some money—nothing seemed to impress her." Then, after a pause, he told me he had realized something: that no matter what happened with the women in his life, he could never prove that his mother was kind and loving to him. And no matter how many women he had, he could never prove to his parents that he was a lovable person. Matthew was beginning to untie the knot of shame.

From then on, within the safe haven of our reflective dialogues, a new kind of self-state seemed to emerge in Matthew. One day he said, "I feel like there is this solid place in me, where I can just watch, just observe, and collect it all." He spoke quietly, and with a kind of surprise at his discovery, which I took in with my own feeling of gratitude.

FINDING A CORE

Is there any core self beneath all our layers of adaptation and personality? I've talked about our multiple self-states, each carrying out its own mission to fulfill our motivational drives: for connection, for creativity, for comfort. Other states coalesce around specific activities: our expertise at a particular sport, our mastery of a musical instrument, or a set of skills necessary for work or school. Still other states operate in our social roles: We lead a community organization, find a romantic partner, participate in family life, make new friends and keep up with old ones.

But beneath all these self-states, I believe, is a core self that has receptivity at its heart. Some researchers call this core *ipseity,* from the Latin word *ipse,* meaning "itself." Ipseity is our "suchness," the being that underlies the activity of each of our self-states. For many of us, this receptive self is hard to imagine, much less feel. But it is the essential "you" beneath narrative and memory, emotional reactivity and habit. It is from this place that we may suspend the flow of top-down influences and come close to what has been called "beginner's mind." When Matthew said that he had found "this solid place" within himself he was describing the receptive self that rests beneath each of the many self-states that are activated in our day-to-day lives. If he continued to develop it, this self could become an inner sanctuary for him, open to what is, ready to receive whatever arrives at the door, inviting all aspects of himself into the shelter of his receptive mind.

In my own and many others' experience, developing mindsight's lens can give us access to this receptive self beneath our layers of adaptation, even beyond our state of mind in the moment. When we develop the spaciousness of a receptive mind, we come to see mental activities, including states of mind, as just the activities of the mind, not the totality of who we are. Resting in the "hub" of the mind, we can achieve a sense of our receptive self, opening ourselves to a world of new possibilities and creating the underlying condition for state integration.

WE-STATES OF JOINING

As time went on, Matthew began to date a different kind of woman—women he found interesting and attractive, not just challenging. Initially his feeling of wanting the "chase" would take him over, and he'd feel uneasy and "bored" whenever anyone liked him. But over the many months of his inner work in therapy, he came to feel differently. Joining became the new focus of his relationships, rather than the old drive to seduce and conquer. Eventually he did find one particular person to care about, and he is now learning to live with the uncertainty that being close entails. Becoming part of a "we" is pushing Matthew to experience his vulnerability and to stay present rather than to fight, flee, or freeze. His sense of shame still emerges in various situations, but now he usually notices his reactivity before he starts to act on it. His newly developed skills have freed him from his automatic patterns so that he can choose how to respond. He and I both believe that he can now create the loving connection with a woman—and with himself—that he deserved so long ago.

11

The Neurobiology of "We"
Becoming Advocates for One Another

DENISE WALKED INTO MY office with a stride that declared security and assurance. She was shadowed by her husband Peter, whose slow shuffle and downward glance radiated dejection as they came in for their first couple's session. Denise sat upright in her chair; Peter slouched on the couch, immediately picking up a large pillow to hold in his lap as a kind of shield. You didn't have to be a psychiatrist to know they were a couple in trouble.

"He's a wimp," Denise declared. "And what's more, his neediness makes me sick!"

Peter seemed out of breath as he spoke, but that didn't prevent his own attack. "It doesn't take long to figure out this isn't working. I married a narcissist. What was wrong with me?"

You might reasonably imagine that with so much hostility and contempt right out of the gate, this relationship was beyond repair. But beneath this pair's rage and disillusionment I sensed sadness and loneliness—and perhaps a longing that might motivate them to change.

Denise and Peter had been married for ten years, and they had two young children at home—whom they said they each loved but relentlessly fought over. Both of them were in their late thirties and were involved in intense professional careers, Denise as an architect, Peter as a teacher at a leading conservatory of music who also performed occasionally. They had tried marriage counseling for a while, but found that their efforts to "open up the lines of communication" were futile. Their next appointments, Denise said, were going to be with their attorneys. But they felt they owed it to the children to try one more time, and a friend had suggested they call me.

Denise continued to spell out her complaints. Early in their marriage, she'd felt everything was "fine," but as the years had gone by, she had come to see Peter as "very insecure and too demanding." Her emphasis and conviction created an image in my mind like a neon sign: HE IS ILL AND NEEDS HELP. She'd always known Peter was "an emotional guy," she said, but she hadn't realized until they had children that he was actually "weak." She told me that he couldn't, or wouldn't, stand up to their two-year-old daughter, who had him "wrapped around her little finger"; he put up with tantrums that Denise had "no time for." He was no better with their five-year-old son, she complained. "He sweet-talks and negotiates and lectures, and the boy just ignores him. He should just tell them both to shut up and do what they're told!" she concluded. "I've lost any respect I might ever have had for him."

Peter's concerns revolved around feeling isolated in the marriage. "Denise is too independent and strong-willed. She's hard on the kids, and she's hard on me. I never see her warm up to them—she's like a boss without a heart." Peter went on to say that he felt lonely and uncared for by her. As he told me this, he looked away from me, and from Denise. He sounded forlorn and helpless.

LIVES OUT OF HARMONY

The brain is a social organ, and our relationships with one another are not a luxury but an essential nutrient for our survival. But both Denise and Peter were in profound distress. Clearly, the way they were relating to each other was anything but a source of well-being.

What could we hope to achieve in therapy? Could either of them change enough, individually or as a couple, to bring their relationship back into harmony? Sometimes the best a couple's therapist can do is help two people see how mismatched they are, so that they can separate and move on. Clearly Denise and Peter no longer shared the experience of "feeling felt" by one another, if they ever had. The sensation of being with someone who knows you, who wants to connect, who has your best interest in mind—this essential nutrient was missing from their relationship.

I asked to meet with each of them individually before coming up with a plan for therapy. In these visits, I confirmed that they both genuinely hoped the marriage could be saved. There were no infidelities, no betrayals, no secret agendas or hidden convictions that the marriage was dead and beyond repair. When I'd seen Denise and Peter together, they'd had flashes of the kind of mutual contempt and vindictiveness that can doom therapy. But when they were alone with me, the longing that I'd sensed beneath the surface in our first meeting emerged; they weren't just seeing me "for the sake of the children." Peter sounded less resigned and negative. He talked about his respect for Denise's strengths, and how at one time they had "made a good team." Denise was initially more aloof, but as we continued to talk she seemed to soften. Unlike the faultfinding stance she'd taken in their joint session, she told me that she wanted to see what she could do to make things better. That was a surprise, and further gave me a sense of hope. Even if they didn't, in the end, stay together, I could at least help them part amicably so that they could coparent with minimal animosity.

And so I said that I'd be happy to work with them, and they agreed to a limited number of sessions. After six meetings we would reassess where we were and make a collective decision on how to proceed. As a first step, I felt that I could draw on the positive intention they expressed when alone with me to help them shift out of their defensive and reactive patterns and move towards openness and vulnerability with each other.

Ironically, the aspects of a person that we find most attractive at the beginning of a relationship become the very characteristics that drive us mad later on. In our next joint session, I asked them about the beginning of their relationship. Peter said that he had been drawn to Denise's "independence, strength, and strong opinions" and thought that they were a good "complement" to what he felt lacking in himself. Denise said that she was at first attracted to Peter's "appearance, sensitivity, and the way he talked about his feelings." She didn't know exactly why these were traits she liked, but she "just did." Peter looked surprised, even hopeful, when Denise said this. But then she went on to repeat that she now found him "too emotional" and "extremely insecure." Her judgmental tone wiped the look of openness right off Peter's face.

In the voyage from romance to marriage, something had changed. They both became busy in their professions, and their relationship went on the back burner. Time moved forward, children came, and they found themselves becoming irritated with each other, frequently, and with a surprising intensity.

Peter described a typical conflict: He'd "want to be close" to Denise, to talk over his day with her or "just get a hug" when he came home from work. But she'd either be preoccupied with the children's routines or she'd "just pull away" from him and withdraw into her home office to seek solitude. Peter's response to her withdrawal was to reach out with more intensity. "I just can't stand it when she shuts me out like that," Peter said. (Denise's face looked blank when he said this.) But if he protested, Denise would yell at him and tell him he was too demanding. Now, he said, he'd begun to doubt his own feelings. Did he have a right to want to be close to his wife—or to anyone?

Over time this pattern of his approach and her withdrawal had evolved into a distancing set of interactions. They couldn't pinpoint any particular disagreement or event that marked the beginning of their troubles, but, Peter said, their relationship was starting to feel dead even before their daughter, Carrie, was born. While Peter felt that he was "dying on the vine," Denise initially said they could "survive" their conflicts if "he'd just leave me alone." Their sexual life had dropped to near zero in the past year, and Denise said that this was "just fine" with her. "It isn't fine with me," Peter shot back. I also learned that Denise had recommended early on in their troubles that Peter seek out his own therapy, which he had done, but nothing seemed to change—in him, or in her. While each of them may indeed have needed some individual work, the "we" of Denise and Peter was in desperate need of attention.

There was more behind this pattern of interaction than the "communication problems" they'd tried to address in counseling. Denise and Peter actually communicated fairly well, at least on the surface. Both of them were quite articulate, and they even listened to each other—they'd grasped the basics. But kindness and compassion were in short supply in the marriage. Denise and Peter talked about each other largely as a collection of annoying, hurtful, or inadequate behaviors. Neither expressed much respect for the other's mind or much interest in the internal experiences of the other. This lack of insight and empathy was keeping them from finding common ground from which to address their differences.

FEELING SAFE WITH EACH OTHER:
RECEPTIVITY AND REACTIVITY

A mindsight approach to couple's therapy differs from other strategies in that we pay close and careful attention to the flow of energy and information—how it is regulated by the mind, shaped by the brain, and shared in our relationships. It was time to introduce Denise and Peter to the triangle of well-being and to the notion of integration. When I showed them the hand model of the brain, I especially emphasized the way the brain creates the two distinct states of mind I'd observed in them both, so that they could grasp the fundamental difference between an open, receptive state and a closed, reactive one.

In order to help them experience this difference directly, I offered them a simple exercise. I told them I would repeat a word several times and asked them to just notice what it felt like in their bodies. The first word was *no,* said firmly and slightly harshly seven times, with about two seconds between each *no.* Then, after another pause, I said a clear but somewhat more gentle *yes* seven times. Denise reported afterward that *no* felt "stifling—it really pissed me off." Peter said it made him feel "shut-down and tight, like I was being scolded." In contrast, the *yes* made him feel "calm, with a peaceful feeling inside." Denise said, "I was glad when you started saying 'yes,' but I was still mad from the 'no' thing. It took me a while to relax and feel okay."

Now that they had an immediate felt experience of the difference between reactive and receptive states, I went on to explain that when the nervous system is reactive, it is actually in a fight-flight-freeze response state, from which it is just not possible to connect with another person. I pointed to the palm of my hand model of the brain and explained how automatically and swiftly the brainstem reacts whenever we feel threatened, either physically or emotionally.

When our entire focus is on self-defense, no matter what we do, I said, we can't open ourselves enough to hear our partner's words accurately. Our state of mind can turn even neutral comments into fighting words, distorting what we hear to fit what we fear.

On the other hand, I went on, when we're receptive a different branch of the brainstem system becomes activated. Their responses to the *yes* had already suggested what happens: The muscles of the face and the vocal cords relax, blood pressure and heart rate normalize, and we become more open to experiencing whatever the other person wants to express. A receptive state turns on the social engagement system that connects us to others.

In a nutshell, receptivity is our experience of being safe and seen; reactivity is our fight-flight-freeze survival reflex. My first challenge to Denise and Peter was to ask them to just notice the state they were in—or the one that was emerging—as they began a discussion. If either of them was in, or anywhere near, a reactive mode, they should stop and ask to take "a break" or a "time-out," which the other person would agree to respect. For now they could take as much time as they needed to calm down—as long as they both agreed to come back to the communication table when they were ready.

As our sessions progressed, Denise and Peter began to recognize what these states felt like in real-time interactions. At first I'd be the one to pause when I felt one or the other of them was getting reactive, and I'd raise my hand as our shared signal that it might be time to take a break. Soon they each learned to detect this reactive feeling inside—when they'd slip away from receptivity—and they began to initiate the pause that refreshes on their own. They were a bit surprised how hard it was to ask for a pause when the other was speaking, and even harder to accept when they were the one talking. At one point, Peter said that Denise's pause signal felt like she was saying "just shut up" (at this Denise scowled)—but then he went on. He said he'd just realized she was telling herself to stop, too. At this Denise's tight face loosened, and I could see her eyes soften a bit, as if she had just discovered something hidden and important. Then she reassured him, with a slight smile, that she'd tell him directly to "cram it" if that's what she meant. This brief interaction, ending with a touch of humor, was a hopeful sign. Peter could learn to recognize and adjust his perceptions; Denise could acknowledge that he'd done so and show a little perspective on her own behavior. It was a small moment of joining and collaboration.

In a later session, Peter once told Denise that she was showing her "narcissistic side again"; he spoke quietly, but it wasn't hard to pick up his anger and his intent to insult her. Previously, Denise would have lobbed an insult right back, with Peter's emotional "neediness" as the easy target, but instead she raised her hand. "I'm going reactive and we have to stop." They both paused and focused on their breath. I wish I'd had a camera in the room to show you how this transaction unfolded. After their pause, Peter was able to acknowledge that he'd lashed out in fear. This allowed Denise to own her accurate perception of his intent—and to forgive him for the attack. What would once have driven the nails deeper into the coffin of their relationship now became an occasion for repair and rebuilding their trust.

OPENING MINDSIGHT'S LENS

Denise and Peter had spent so much time together in reactivity that they needed to strengthen their basic ability to enter a receptive state of awareness. To help them unlearn the old pattern and learn the new, I spent much of our third session introducing them to the hub of the mind and to the basic mindfulness-of-the-breath exercises. I used the hand model of the brain to explain how focused attention could help develop the middle prefrontal areas, and I also explained how this would support what we were doing in therapy.

Peter had done some yoga when he was younger, and he immediately found the basic practices quite calming. But staying present in this way was new for Denise, and she told me she found the exercises odd, unhelpful, and somewhat confusing. I encouraged her to just notice that odd feeling and not to expect the practice to do anything in particular. To her credit, Denise stuck with it and did the exercises at home, but it was quite a while before she could open to any sense of clarity or calm.

By now, of course, you know that my goal for these integration-of-consciousness exercises went beyond calm. I wanted to give

Denise and Peter a way to increase their capacity to find that core place beneath their individual adaptations, the receptive state buried under layers of reactive defense. In the case of Jonathon in chapter 5, reinforcing the prefrontal circuits enabled him to pause and avoid being swept up by his wild swings in mood and helped to stabilize his roller-coaster mind. I had a similar hope for Denise and Peter: that widening the middle prefrontal hub would enable them to see beyond their reactions and so find each other.

I also trusted that it would help them find themselves.

MAKING SENSE OF THE PAST TO FREE THE PRESENT

For our fourth and fifth sessions, I decided to conduct an attachment interview with Denise and Peter as each listened in to the other's unfolding story. I asked them directly if they would honor the vulnerability this involved. They agreed, both in words and with nonverbal signals I could clearly feel, that they would respect the inner world of the other as it emerged in the interviews. This agreement, plus the basic goodwill they had shown in their individual sessions, made me feel we could do this work respectfully.

What emerged, in brief, was that Peter had a generally preoccupied narrative that revealed his continuing concern with leftover issues from his childhood, while Denise had a dismissing narrative that minimized her need for others—both then, in her childhood, and now.

Peter was the youngest of four siblings, and his mother had developed chronic back problems from a car accident shortly after he was born. There were a number of surgeries, hospitalizations, and long periods of convalescence at home, while Peter's father worked two jobs as a security guard, including some night shifts, to make ends meet. Peter's oldest sister, Maggie, who was a dozen years his senior, was the primary caregiver for Peter during his early years, but she developed a drug addiction as a teenager. (She'd started with their mother's pain medications and moved on to barbiturates and alcohol.) Maggie would leave him alone with his other sisters, five and seven years older, who, Peter said, "just fended for themselves."

"I would try to find my way to my mother," he recalled, "and sometimes it would be fine—I mean, she'd be there for me. We were close for a while—when I was young, I think. I know she used to spend a lot of time with Maggie—she liked her more than my other sisters or me. But mostly she was either locked away in her room or didn't seem to care. She never has. And I am still alone," Peter concluded somewhat bitterly. This intermingling of the past ("I would try to find . . .") and present ("She never has. And I am still alone") not only revealed a preoccupied state of mind, but also suggested something about how Peter saw Denise.

Throughout Peter's childhood, his father smoked two packs of cigarettes a day "to relieve his stress," and he died of a heart attack when Peter was fourteen. His mother recovered somewhat after that and began to work as a supply teacher, but Peter never regained a sense of closeness to her. She was "a sad, depressed woman who remained alone" for the remainder of her life. (She'd died ten years earlier, shortly before his marriage to Denise.) Peter said that he'd always felt responsible for her sadness, especially during the years when he was the last child left at home. Music became Peter's refuge. His talent won him the praise he'd seldom received at home and gave him a way to vent his creative energies. His music also landed him a scholarship at a conservatory across the country.

When Peter went off to school, he was determined to remain financially independent so that "I wouldn't need to rely on anyone for anything." He lost touch with his siblings and made what he called "duty visits" to his mother only once or twice a year. He did well in his studies, discovered a passion for jazz, and was admitted to a top graduate school in musicology, but he found himself in romantic relationships with women who asked "too much of him" and who never helped him feel "at ease." He could never make them happy, he said. He was certain that they'd leave him, fearful that they wouldn't. These stormy relationships made him moody, irritable, and "unstable," and he'd "fly off the handle" quite frequently. Even his performance as a jazz pianist began to suffer. "I couldn't get into that 'head space' to just let the improvisations happen. I actually thought of going back to classical pieces I could sight-read off a score." Then, during his last year of graduate school, Peter met Denise at a friend's party, and he felt "safe" with her. She never asked much from him, and he was relieved to find that there was "room in their relationship for me to feel comfortable and to be myself." His jazz performances improved, and things seemed to be on the right track during the first years of their marriage.

Denise's story was very different. Her parents had been in good health, and there were "no particular issues" that she could recall. In fact, she said that she didn't remember many details about her childhood, except that it was "average, what a normal childhood is like." You may recall that this kind of vague summarizing and glossing over specific details is characteristic of dismissing narratives. When I asked her more directly about her relationship with her parents—what would happen when she was upset or when she was separated from them—Denise responded: "My mother took good care of me. She was neat and an excellent cook. There were no particular things—I mean nothing that would get me upset. My father was the same way. He was an engineer. My mother worked as a secretary, and we had a very organized home. Not that it had to be that way. We chose it." Notice that the question is about "relationships," but Denise's responses focus on the individuals involved—a common pattern for someone with a history of avoidance and the adult dismissing stance.

Then we came to the attachment interview question about loss. "Yes," Denise replied. "There was a loss when I was a child. My brother developed leukemia when I was seven years old. He was only two, and I don't remember much except that we didn't talk about it after he died. My parents just went on. It didn't seem to change that much, I think. Now there were just three of us again." Denise told me in a rather neutral way that she sometimes wondered why no one ever talked about her brother's death. I made a few more attempts to explore the family's emotional response to this loss, but Denise continued to deflect the conversation.

Despite her "relationships don't matter" stance, I hoped that Denise's fundamental need for connection was intact, and I believed that she could become more aware of this need if we approached it cautiously. As I've mentioned, research has revealed that people with dismissing narratives show physiological signs that their subcortical limbic and brainstem areas still register the importance of relation-ships. It's simply that the higher cortical areas, where consciousness is created, shut out this awareness in order to survive barren times. The key would be to align myself with these deeper subterranean circuits and bolster Denise's ability to integrate them into her life.

At the end of Denise's interview, I came back to her brother and said that perhaps feeling her feelings hadn't been safe in a family where everything was so "organized and neat," and where people didn't let themselves respond to the death of a child. She just looked at me with wide-open eyes. This was very different from the confi-dent gaze she'd worn when she first strode into my office. But no words came. Not yet. She just looked at me, and my mind noted that something had shifted inside her—something that had to be respected, not inspected, and not addressed overtly at this vulnerable stage of her internal unfolding.

Both Denise and Peter had done the best they could to survive dif-ficult childhoods, and their adaptations had left each with a gap in their development that the other person magically filled when they first met. We all long—consciously or not—for what we did not receive in the past and what we don't have now. Denise could have used some of Peter's access to his feelings and his ability to be spon-taneous and connected to his internal world. Peter could have used some of Denise's capacity to distance herself from her emotions and needs, to stand back a bit from troubling experiences. But instead of collaborating and learning from each other, they had retreated to their respective extremes as so often happens with cou-ples in distress. Now they were stuck at these distant poles.

THE DECISION TO CHANGE

Picture how Denise's mind had sculpted her brain so that she could survive growing up in an emotional desert. In response to an avoidant attachment to both parents (the most likely explanation for her dismissing narrative), she had shut down the circuitry of her brain that needed closeness and connection: the relational, emo-tional, and somatic-focused right hemisphere. This is how she had become a "boss without a heart"; she was disconnected from her own internal world of feelings and bodily sensations. Like Stuart, she seemed to have sought refuge in the logical, linear, linguistic, literal left mode of living. And like Anne, who had cut herself off from the neck down, Denise also seemed quite disconnected from her subcortical world. Even in her work as an architect she had gravitated to office and industrial design, rather than to homes or gathering places such as libraries, schools, or museums.

The question now was whether Denise wanted to keep up this emotionally disconnected way of being. I thought I could help her by starting from the safe distance of science and by appealing to her highly visual architect's mind. Using a detailed larger-than-life plastic model of the brain, I showed her the hemispheres, pointing out how the corpus callosum linked the two halves, and explaining how these connections may have been shut down. I also lent her a book for parents in which I'd described these brain adaptations. Once these synaptic realities became clear in her awareness, I also reminded her that the brain is open to change across the lifespan. Because it responds to the focus of attention and to the experiences we create intentionally, there was great hope that those unrealized neural connections could still be stimulated to develop.

I phrased this idea as an invitation from me and as an opportunity to grow, not as something she had to do to meet Peter's demands. This was a crucial point. Some people's attachment histories make them conform too readily to others' expectations, while others shut off any input from people close to them. These conditions can distort our motivation to change. Inviting collaboration works better than issuing an ultimatum—"change, or else." Denise could elect to stay the same, I said. It was ultimately her choice, and I asked her to think about it during the following week before committing to a certain path.

Peter, for his part, needed to know how the wash of feelings that flooded him at times related to his early relationship experiences. Being ignored in any way, not being "seen," was a powerful trigger for him. Sometimes he'd jump suddenly from negotiating with the children to yelling at them. If Denise cut short a discussion when they had a disagreement, he would sulk around the house. But if she retreated to her office and locked the door, he might "lose it" and start banging to be let in. And at the music school where he taught, he might explode with frustration if he wasn't consulted about a change in his students' scheduled lessons. (The director "ignored and disrespected him," he said.) These low-road reactions suggested that his prefrontal cortex was vulnerable to going off-line, and that his right-hemisphere processes could overwhelm the balancing influence of the left. In effect, Peter tended towards chaos while Denise tended towards rigidity. They were trapped on opposite banks of the river of integration.

If, like Denise, you've spent your childhood in an emotional desert, linking to others would yield no nourishing attunement, and in fact might be dysregulating. Your windows of tolerance for mutual connection, much less deep intimacy, would be quite narrow. One strategy would be to adapt by shutting down the circuit that reminded you repeatedly of what was missing. If Denise's parents never spoke of their feelings after the death of her brother, if they never even acknowledged the loss, it's unlikely that Denise would have felt safe in any part of her internal world. There would have been little joining, few moments of the resonance that creates a "we," little closeness to ease her sense of isolation.

Peter's childhood history with a responsible but absent father and inconsistent caregiving from both his mother and his sister had shaped his resonance circuitry in a different way. Support was intermittent, and both his sister's addiction and his mother's reliance on pain medication had probably created times of overwhelming and unpredictable communication. His mother had been emotionally blunted by depression as well. Growing up with a depressed mother has been shown to significantly affect the developing brain of the child; for the child, it's like living in a chronic "still-face" experiment. During his early years, Peter would have needed to amplify his attachment circuits to attempt *any* kind of connection, and his window of tolerance for disconnection would have narrowed, so that at any hint of rejection he'd withdraw in misery or erupt in chaotic anger.

I took care to explain to Denise and Peter that although his present state of mind *was* in part a product of his difficult past, the despair and loneliness Peter was experiencing now were responses to real and ongoing experiences with Denise in the present. However, the pattern of response that had been shaped in his childhood had helped to fuel the ongoing negative feedback loop in which he and Denise were trapped, with Denise's reactive withdrawal from his neediness pushing him even further into isolation and hopelessness. We could focus our work on their current lives, and the past would come with us, ready to be illuminated and explored.

The question now was whether he could calm his anguish enough to help create a healing space for each of them, so that they could achieve more connection both in their individual minds and in their relationship. Breaking their cycle of agitation and disconnection was not the responsibility of either alone. Somehow the dyadic system, the dysfunctional way this couple acted together, had to be changed.

Peter's choice was this: If he wished to move forward with Denise as his partner, he would need to learn to regulate his internal distress more effectively so that a broad range of responses from Denise would not throw him into a tailspin. Denise's own work would involve learning to become more connected—first to her own body and emotions, and then to Peter's internal world. She would also be challenged to widen her window for taking in Peter's signals so that his needs wouldn't automatically push her into retreat. I described these as two "growth edges" that might be the first focus of a longer therapy.

After a week away for reflection, both Peter and Denise elected to continue our work together past our original six-week starting point.

DISTORTED MIRRORS

The morning before our next session, I awoke with the tune of one of my favorite songs, James Taylor's "Carolina in My Mind," playing in my head. It had acquired some new words:

> In my mind I'm driven by mirror neurons.
> Can't you just see intention
> Can't you just feel emotion.
> Ain't it just like history to sneak up from behind
> 'Cause I'm driven by mirror neurons in my mind.

> There's a holy host of others gathered between us.
> Maybe we're on the dark side of the road
> And it seems like it goes on and on forever.
> You must forgive me
> 'Cause in my mind I'm driven by mirror neurons.

I wasn't about to share my musical musings with Denise and Peter, but those words reminded me of where we needed to go.

As you may recall from "Riding the Resonance Circuits" (page 59), mirror neurons are the antennae that pick up information about the intentions and feelings of others, and they create in us both emotional resonance and behavioral imitation. We engage in mirroring automatically and spontaneously, without conscious effort or intention. In my mind, the "host of others" that put us on "the dark side of the road" are the suboptimal influences of our early relationships that dim or distort our mirrors.

Our mirror neuron system "learns" by how it couples our own internal state with what we see in someone else. After explaining this to Denise and Peter, I asked them to consider how their individual lives in the past might have created the reactivity they experienced with each other. It was striking, I said, that when I'd interviewed them alone, each was not only receptive to me but also open to the virtues of the other. Yet when they were together, everything seemed to fall apart. "Maybe we should just live in separate houses?" Denise quipped. She was smiling fully for the first time in our three-way sessions.

Her use of *we* was also a good sign. I'd noticed that she rarely used it in talking about Peter and herself, and I wondered if she ever thought in those terms. Avoidant attachment creates an impaired sense of the importance of joining, with a blocked sharing of right-hemisphere signals. These are the very signals the mirror neuron system uses to simulate the other within ourselves and to construct the neural map of our interdependent sense of a "self." It's how we can be both an "I" and part of an "us." Denise didn't seem to have developed much of this capacity.

Now consider Peter's mirror neurons. As a baby he is born ready to connect, ready to link what he sees in others with what he does and with what he feels inside. But what if those others are some-times attuned and clear, but more often not available, and at still other times intrusive and bewildering? Unreliability and confusion come to permeate what Peter sees in others' minds; they also shape how he experiences his own mind, and even how he creates and knows himself. He has, in simple terms, a "confused" inner self.

Peter's "preoccupied" adult narrative suggests an ambivalent attachment as a child and a consequent amping up of his attachment system. He would have been thrown repeatedly into a reactive state of alarm. Is my attachment figure here to soothe and protect me? Can I depend on her to see me and keep me safe? These experiences would have primed Peter's brain to be especially challenged by his mother's depression and her unresolved grief after his father's death, none of which he could heal, as he'd once said, "no matter how hard I practiced and performed, no matter how hard I tried." They also became a part of his implicit memories, free-floating states of mind from the past that were capable of shaping his state of mind in the present.

Studies have demonstrated that the brains of people with a his-tory of ambivalent attachment are in fact more sensitive to negative feedback: Their limbic system's amygdala fires off more readily in response to angry faces. Peter's sensitivity to Denise's hostility and rejection fit this picture well. The same studies revealed that in peo-ple with avoidant histories, the circuits for social reward are actually damped down: They respond less to smiling faces. This would help to explain Denise's resistance even to positive attempts to connect on Peter's part. If Denise and Peter could see their differences as based partly in differing brain sensitivities, they might be able to move beyond their habit of mutual blame.

ENTER, THE BODY

I knew these ideas were a lot for Denise and Peter to take in, but I wanted the framework of our work to be clear: Neither was to blame for their current situation—and they would need each other to support their personal and relational growth. In addition, to truly take in the other, they needed to be open to all the subcortical sensations—from the limbic areas, the brainstem, and the entire body—that would create and reveal resonance. Before they could become a "we," I had to help both of them find the connection to their own bodies—which I would do by teaching each of them the body scan.

In our next session—with all of this in my mind—I felt deeply moved when I observed Denise and Peter performing the body scan, together, with focus and intention. When they both emerged from the internal practice, there was a calm in the room that I could feel but can't really describe. Their faces looked softer, the tone of Denise's voice felt more relaxed, the worry on Peter's face had melted away. There was an openness, even after this first practice, that I think they both could also feel. I didn't say much about it, but the three of us seemed to be breathing sighs of relief.

In subsequent sessions, a brief body scan would become a place for us to start—or to turn to when they needed a pause from an interaction that was becoming reactive. Grounding themselves in this inner world of the body created a safe place for both of them.

DIFFERENTIATION AND LINKAGE

Attuned couples link together in a mental lovemaking, a joining of minds, in which two people create that beautiful resonant sense of becoming a "we." The intimacy that blossoms can be amazing, but the journey to get there and remain there can be rough. To become linked as a "we," a couple needs also to become differentiated as two "me's."

Denise's growth edge required that she widen the hub of her mind to enable her to become aware—safely and slowly—of bodily feelings and limbic emotional states that had previously eluded the radar of her conscious attention. I taught her a simple process for deliberately accessing the Sensations, Images, Feelings, and Thoughts (SIFT) that she had previously shut out. Using SIFT as her checklist, I asked her to sort through her responses to ordinary daily events. Denise was usually aware of her thoughts—she was comfortable in her logical left hemisphere. But there was a whole new world she needed to explore: the sensations, images, and feelings of her right hemisphere, including the deep longings that had been hidden away since childhood. SIFTing her experiences was a gradual, safe way to begin this exploration.

Peter worked on his own growth edge to widen his windows of tolerance for being alone. As he became more attuned to his body, he used his left hemisphere's emerging capacity to "name it to tame it." When Denise needed time to go inward and work things out during a discussion, he'd become aware of his heart starting to pound, his jaw clenching, and his hands tightening into fists. Then he'd use the mental notations he'd learned in mindfulness practice—"Anger" or "Frustration" or "Despair"—to label his feelings. He found that if he paused, the rush of feeling would just arise and fall in the space of his mind. As he'd learned to say, "A feeling is not a fact."

Harnessing the hub of his mind, Peter could now use the strength of his left hemisphere to label and describe, to approach and not withdraw. He still had all the power of his right hemisphere—the internal feelings and maps of his body were fully present—but his inner world was no longer throwing him into chaos.

Once Denise and Peter had begun to create more integration within themselves, I felt they were ready to focus more directly on their relationship with each other. I wanted to help each of them sense the mind of the other with respect, to be able to share their memories and narratives. Then they could, as a couple, make sense of their individual past experiences and understand how their history together had been molded, in part, by the adaptations they had made to survive their childhoods. Now together they could help themselves discover the world of we.

I challenged each of them to become an expert in sensing and respecting the other's mind: With the curiosity, openness, and acceptance that create love, each could approach the other as both interpreter and supporter. "The job for each of you," I said, "is to become the advocate of the internal world of the other."

With this assignment, Denise would try to read the nonverbal cues coming from Peter and let her own body be receptive to whatever she was sensing in him. She could let her mirror neuron system just soak in his signals and shift her own internal state. At the beginning of our work, Denise would probably have checked out at the very thought of this assignment. But by this time she was able to not only stay receptive, but also to become an active advocate for Peter.

One day when the couple came in, Peter told this story. Two days earlier he had learned that a new instructor was being promoted to head the piano section at his conservatory. Peter had seniority, and he'd wanted the job. Now this new man would be his boss. In the past, Denise had expressed a lot of opinions about Peter's "mild manners" at work: She'd pointed out repeatedly that they kept him from asserting himself or getting what she thought he deserved. But now Denise realized that being Peter's advocate was not only a matter of recognizing his musical talent or defending his right to a promotion. This time, she was able to simply stay open to his feelings of disappointment when he came home. "She asked me about how I felt," Peter said, "and she wanted to know all the details about how I heard the news and how I responded. You know," he went on, "in the past she would have just complained that I was a wimp—and told me what I should have done. I couldn't believe how she was just listening to me. It felt good."

I thought for an instant that Denise might focus on the "wimp" issue and be offended, but she was smiling. She spoke right to Peter. "You know," she said, "I could feel how down you were the minute you walked in the door, and then when you told me about that guy, I knew it must have been hard for you. I thought about Maggie—about how your mother favored her—and about how the director just did the same to you." Instead of adding to Peter's humiliation, Denise had aligned herself with his internal world—and become his true advocate.

In his own words, Peter was "blown away" by the idea that Denise could actually *see* him and defend his right to his own feelings. Peter in turn was consciously trying to respect Denise's need for a bit more distance, especially at times of stress. This was new for Peter, holding Denise's needs in mind without impulsively expressing his frustration when she did not immediately meet his desire for closeness. He was learning the power of cortical override—to squirt that "GABA-goo" from his middle prefrontal region down to his irritated amygdala to soothe its firing. Each of them was intentionally stretching their early adaptations to meet the other somewhere in the middle.

I hoped that the goodwill embodied in these intentions would activate feelings of safety in both Denise and Peter, allowing their "alert modes" to relax into a feeling of connection and openness.

It was as if their middle prefrontal regions at the top of their resonance circuitry were learning a new way of being. The key was not to take their own or the other's old patterns of response so seriously, or so personally. Simply put, reactivity cuts off seeing clearly. They had to unlearn those old automatic responses before they could create a new, more receptive state together.

As Denise came to appreciate Peter's efforts and his newfound ability to give her "space," her own growth edge moved. She began to notice new inner sensations—a tightness in her throat, heaviness in her chest, an empty feeling in her belly. She was learning to "just let these sensations be" instead of shoving them away. Sometimes she'd know what they meant, but often she simply had to sit with them. She said that she was beginning to trust her body to let her know what mattered. "Even if my head tells me that nothing is going on, now I can feel something, like an internal alarm, that tells me the truth."

As our sessions went on, Denise came to feel more open to these feelings and to her need to be close to Peter. Curiosity about the internal world, her own and Peter's, was her starting place. We explored how, as a child, she'd never been given the gift of being seen by others, or supported in how she felt. This was something she now shared with Peter. She started to recall a bit about the loneliness and fear she'd felt when her brother died, and the strange silence that had filled their house afterward. These weren't dramatic revelations—people with avoidant histories just don't have vivid autobiographical memories to return to. Yet now Denise the mother could mobilize her imagination and empathy for the little girl she'd once been. No tears flowed from her eyes, but at that moment her vulnerability filled the space between us.

To be with Denise in the room that day was like seeing a new way of life open up for her. Something positive had begun to grow in her—a feeling of energy you could sense, a new generosity towards herself, towards Peter, and towards her children.

A "WE WHEEL OF AWARENESS"

In the months that followed, Denise and Peter embraced the ebb and flow of their progress. Although when we first began they'd both doubted that change was possible, they were beginning to see the fruits of their labors. We'd use part of each session to explore issues that had arisen during the week—disagreements about how to handle the kids, misunderstandings about social plans, things that triggered their old patterns of approach and withdrawal. Woven into our work was continual attention to the unfolding of their shared narrative, connecting present experience with meanings from the past so that challenges became opportunities for further growth.

One day, Denise wanted to talk about a night the week before when she'd had to stay at the office very late to finish up a huge project. She'd told Peter this was happening, but he'd forgotten, and he became angry when she didn't turn up for dinner. He called her at the office, very irritated, and Denise in turn reacted to his forgetfulness—he hadn't registered this big project, which was so important to her. "I told him I'd be late, and he just didn't listen to me," she said. But that night, instead of hashing over their grievances, they decided to focus on what *the other* had felt.

"I couldn't believe how Denise was when she came home," Peter said, looking over at her with admiration. "She just came upstairs and said, 'Wow you have the kids in bed already?' and then asked if we could sit down and talk."

"I said I hoped we could just listen to what the other had to say—like we do here—and own how we felt, not point fingers or blame. To tell the truth," Denise continued, "I was surprised the kids weren't still up running around—and really grateful not to have to deal with them right then." (Peter had been working on setting clear limits with the kids and giving them the structure they needed—structure he'd never had himself, and which he initially felt guilty and "unloving" about imposing. It was a new thing for Peter to feel comfortable saying "no" before he reached the boiling point.)

Denise went on. "Peter said that he'd been feeling excited about having dinner with me that night. It was the day after his big concert, and he wanted to talk about all the great feedback he'd been getting. When I wasn't there, it felt like a rejection. In the old days I would have just told him to cram it, but I could feel his sadness, and I listened. The fact is, I did forget—I didn't appreciate how important that ensemble performance was to him. I messed up. And I admitted it."

I could see Denise's openness reflected in Peter's face. Then he said, "You know, the details don't even matter that much. Not like the old days when we'd nitpick every word and see who could come up with the worst insult." Denise reached across the couch and took Peter's hand. "I could understand when Denise told me that she felt hurt that I didn't remember her big project. And mostly I just felt relieved that we were talking—instead of my blowing up or her shutting down." Peter paused for a moment, and then said, "You know, I really do get it that I have a thin-skinned brain, and I can't let it get the best of me."

Peter and Denise were becoming a we. Each had a more curious, open, and accepting stance towards their individual—and now collective—internal worlds. Denise mentioned that she also felt they were becoming more connected to their children. "It sounds kind of funny, but I *do* feel like I'm tuning in to what they feel, and not just reacting to what they do. It's a big difference." Peter just smiled and nodded in agreement.

At the end of that session, Peter helped Denise on with her coat, and I noticed that Denise put her hand on his shoulder as she turned to say good-bye to me. "They" left that day, headed to the home and life they were building together.

This is the essence of mindsight: We must look inward to know our own internal world before we can map clearly the internal state, the mind, of the other. As we grow in our ability to know ourselves, we become receptive to knowing each other. And as a "we" is woven into the neurons of our mirroring brains, even our sense of self is illuminated by the light of our connection. With internal awareness and empathy, self-empowerment and joining, differentiation and linkage, we create harmony within the resonating circuits of our social brains.

12

Time and Tides
Confronting Uncertainty and Mortality

WHEN I WAS IN MY EARLY TEENS, some evenings I'd ride my bike from my home to the beach and wander along the wide strip of sand to the ocean's edge. I'd watch the waves and be filled with wonder—about life, the tides, the sea. The force of the moon beckoning the water, raising it up towards the cliffs, then pulling it back down beyond the rocky pools, back out to sea . . . These tides, I thought, would continue their eternal cycle long after I was gone from this earth.

I wasn't alone in these thoughts as an adolescent. The adolescent brain changes, especially in the prefrontal cortical regions, so that we begin to reflect on the self and on life, on time and mortality, acknowledging the transience of the things around us and of our own existence.

By the age of three or four, children have begun to think in concrete terms about death. They realize that people and pets don't live forever. By then our prefrontal regions have also developed enough to begin to weave our life stories. As we move into our elementary school years, our memories move with us and time becomes embedded into our worldview. In adolescence we enter another phase in our prefrontal capacity to sense time—we begin to dream of the future, to wonder about the meaning of life, and to grapple with the reality of death.

When the human brain evolved enough to represent time, the mind riding along its neural firing patterns was presented with an important challenge. On the one hand we have the cortical propensity to create a sense of continuity and coherence, the drive to create a narrative to connect past, present, and future. These cortical connections weave a sense of certainty, giving us a feeling that we can know and control our lives. Within these firing patterns there is also a drive for permanence, a denial that death means finality. Yet in addition to permitting the mind to create these dreams of permanence, certainty, and immortality, the brain is also an information processor that gives us tools to see reality clearly. The prefrontal cortex enables us to know, though we may not readily accept this knowledge, that life is actually temporary, uncertain, and bounded by birth and death. As Vladimir Nabokov put it in the opening to his memoir, *Speak, Memory*, "The cradle rocks above an abyss, and common sense tells us that our existence is but a brief crack of light between two eternities of darkness."

TRANSIENCE, UNCERTAINTY, MORTALITY

As my own children approached adolescence, they asked if our dogs worried about things like death. I told them that because dogs have no prefrontal cortex to speak of, theirs is a relatively simple set of senses, and a simple life of living in the moment without worries about the future. We now know that some of our fellow mammals, such as elephants, have complex forms of grieving and many more certainly suffer in anticipation of harm. Not being able to enter their internal worlds, we don't know how much they may share our capacity—some might say "burden"—to be able to represent to ourselves complex images of life and death and our journey through time.

While many different animal species have nervous systems that enable *anticipation* of events—for example, learning that a flashing light is associated with a reward in a conditioned learning experiment—*planning* for the future seems to be a prefrontal invention. To create representations that move the imagination into the future is the legacy of prefrontal development. This crowning glory of our prefrontal capacity—to remove itself from today and plan for tomorrow—allows us to build buildings, create educational programs, and fly to the moon. In many ways, the prefrontal region can be called the *cortex humanitatis* in that it is essential for so much that is uniquely human.

As we've seen, much of the brain beneath the cortex is involved in here-and-now bodily and sensory processes, such as digestion and respiration, or taking in data from the outside world. This is the work of our five external senses and of interoception, our sixth bodily sense. When we move forward in the cortex—towards the front knuckles and fingernail regions in our hand model of the brain—we come to a neural capacity to perceive things not directly rooted in the physical world in front of our eyes. This is our seventh sense.

The seventh sense allows us to perceive the mind and to create representations of time, not just feel the passage of our days. It tells us that things die, that nothing lasts forever. Our capacity to perceive patterns teaches us about change and that transience is the law of life. At the same time we are aware of our power to influence the things and people around us, so we try to predict and control, to fill our world with safety and certainty.

Yes, the prefrontal cortex enables our mind to plan, dream, imagine, and reflect—and to continually reinvent itself as life moves forward. It creates the seemingly infinite potential of the human mind. But these capacities come at a price.

THE DEATH OF A PRINCE

When I was fourteen years old I was in charge of the garden in the back of our old Spanish-style, single-story home. We grew tangerines, plums, peaches, and even figs in the blazing Southern California sun. My job was to care for the plants and pick the fruit—and, crucially, to water during the hot months in our desert community. I thought it was a great job.

That spring, though, was particularly rainy, and the strawberry plants were wild with exuberance, sending out streams of runners like an octopus extending its long arms to form new plants. The snails had come in abundance, too, to soak in the moisture and banquet on the strawberry leaves and budding fruit. One evening after school, I fetched the snail bait from the garage and sprinkled it over the vines, hoping to preserve the fruit for the farmers, us humans.

I read the label on the snail-bait box: "Warning! Poison! Keep away from small children and pets." No problem: I was the youngest at home and knew enough to wash my hands. And pets: Our backyard guinea pig farm had been shut down for a few years, and the other animals were in cages inside the house. I was raising Emerson, the young son of my old dog, Prince. Prince was a happy, knee-high, sled-pulling mongrel who had found himself a "wife," a beautiful stray Belgian shepherd and border collie mix. We sold all of their litter of six puppies except Emerson, found a home for the mother, too, and I had my hands full with school, the garden, a huge tank of tropical fish, and two dogs.

But two months earlier, during one of Prince's long excursions— on a route he'd been taking for a decade—a car struck him. A neighbor came to our house, crying, to tell us about the accident. My brother carried Prince home, and we said good-bye to him before he died and his body was taken away. I was still recovering from this loss, keeping his son Emerson close to me whenever I was home. In honor of his father—and perhaps to keep Prince alive in some way—we renamed his handsome, smart, keen-eyed son "Prince Junior."

I distinctly remember reading that snail-bait box and thinking to myself, I'd better tell my parents not to let Prince Junior out later, as they often did. I finished my homework, brushed my teeth, washed my face, and went to bed, my young friend sleeping by my side. When I woke up, Prince Junior was dead.

For the longest time I could not stand to look at myself in the mirror—ashamed of the person I saw staring back. I remembered that night, spreading the bait, reading the box, thinking my thoughts, doing my homework, looking at myself in the mirror as I washed my face. Life felt simple, serene, settled. And then the morning arrived, and I realized that I had poisoned my own best friend. And even worse—something I didn't tell anyone—I had in fact read the label, knew what I needed to do to protect him, but just plain forgot. I got into my homework, lost track of that worry, that precaution, and didn't check to be sure everything that needed to be done had been done.

UNCERTAINTY BY THE SEA

Flash forward eleven years. I am working as a senior medical student in a public health clinic near Rincon, a small town on the northwest coast of Puerto Rico. I've taken courses in primary care medicine and tropical diseases and am now serving as the "doctor" for the indigent population living in this surfer's paradise. I don't surf, but I have taken scuba diving lessons in anticipation of days off exploring the Caribbean reefs and caverns.

But now it's just before lunch after a full morning of seeing patients, and I have a vague, uneasy feeling in my belly. I'm thinking about Pablo, a toddler I'd seen earlier that day with severe ear pain and a fever. After taking the history in my emerging but shaky Spanish and doing his physical exam, I'd checked with my attending, and we prescribed an antibiotic for what we decided was a significant ear infection. (I'd had a lot of them as a child, and I still remembered the pain and fear.) I'd watched Pablo and his mum leave, the prescription in her right hand, Pablo's hand in her left.

Now I don't feel right. I get an image of poisoning Pablo. Had I prescribed the antibiotic correctly? Too much, and it would destroy not only the bad bacteria in his middle ear, but also the delicate hair cells lining the inner ear that allow him to hear. I tell myself that I'm just overconcerned and should let it go. I checked with the attending, wrote out the prescription, and it's all just fine, I say in my head. But the nagging feeling won't go away.

At the front desk I find Pablo's chart and look to see what dose I actually wrote down. I discover that I'd recorded only the type of medication, not the amount. Then I check for his family's phone number and find that they live in a distant part of town that has no phone service. I tell myself again that everything is probably fine. But I can't rest. I head for the beach, but instead of relaxing there with a sandwich, I start the long trek south towards Pablo's neighborhood. The palms are swaying in the easterly winds that so often wrap themselves into hurricanes over this coast. I step over the strewn coconuts, over twisted roots of the palms hugging the sandy shore. I can still remember the smell of the pungent air, mangos hanging from the laden limbs, and the squeals and odor of pigs in the yards of the houses ahead.

I wander through the unmarked streets and ask repeatedly "¿Donde esta la Casa del Rios? ¿Señora Rios vive cerca de aqui?" I have to ask people to speak more slowly—but finally I learn that

Pablo's family lives the next calle down, near an empty lot. When I arrive at the house, Pablo and his mother are just inside the front door—and very surprised to see me. I ask to see the bottle of medication, to check on the dose, I explain.

I knew the amount I should have written, based on Pablo's weight. But there it was on the bottle, my first major mistake in medicine: I had calculated the correct daily dose but I had prescribed the full amount three times a day, instead of in divided doses. One day of overdosage wouldn't have been a problem. But ten days—that would have killed the hair cells, destroying Pablo's hearing forever.

I don't know how I knew. It was a whole-body feeling, a gnawing restlessness in my heart and gut that just would not let go. Something was wrong and I had to find out what it was.

I adjusted the dosage, and when I gave Pablo a hug good-bye and said *adios* to his mum, that something inside me that had needed to check felt deeply satisfied. I had a drive to be certain, perhaps because of Prince Junior's death, perhaps because I was in a new role of responsibility. Our minds wrestle with uncertainty all the time, but now I was entering a profession where my drive to know, to be sure, would be activated day and night. Temporal integration was not a luxury, but rather front and center in the work of caring for others.

Today more than ever, medicine is wrestling with these issues. Computer programs enable us to offer precise, step-by-step checklists to help medical staff perform complex procedures accurately. In some areas these checklists have dramatically reduced human error and the resulting complications, including death. But no matter how many checklists we devise, we need to remain open to the wisdom of our whole selves as well, to listen to the intuition that is also the gift of the prefrontal cortex. We then can not only check, but ultimately feel, with clarity, that we've taken care of what needs to be done.

SEEKING CERTAINTY

Just as the waves in the ocean appear to be rolling in from far away, our mind perceives continuities when they do not exist. We spot a big wave far out at sea and watch it surge towards the shore. But in reality the cresting water we see out there is not the same water that surges up the beach a few minutes later. The continuity of the wave is a mirage.

A wide variety of cognitive experiments suggest that our mental perceptions are constructed from a cortical drive to make disjointed reality into a fluid flow of experience. For example, our eyes blink frequently but our brains adjust to the gap in visual input and construct an unbroken picture. The brain has a bias for making the world appear solid and stable. The same could be said about how we develop a continuous sense of a "self" out of the multiple states I explored in chapter 10. And once we learn about cause and effect as young children, we seek causal links in every sort of experience—even making them up when there are none. This drive for continuity and predictability runs head-on into our awareness of transience and uncertainty. How we resolve the conflict between what is and what we strive for is the essence of temporal integration.

WHAT REALLY MATTERS?

When I was in high school there was a period when I was unable to stop thinking about transience and mortality. I remember phoning a classmate to ask her out on a date, or at least that's what I thought I was doing. "Lauren," I began, "how was your day?" She told me about going to the park with friends after school, then shopping for some new shoes.

"And Danny, what did you do after school?" she asked.

"Well," I said, not being one for beating around the bush, "I was thinking about how one day none of us will exist. I just can't get how we are supposed to take things so seriously—like homework and grades and are we going to win the championship. We're here right now, but one day we'll be gone."

There was silence at the other end of the phone. "Lauren . . . are you there?" When I heard the click as she hung up, I knew I was alone with my worries again.

Ultimately, wrestling with transience and mortality requires that we dive beneath the illusion of permanence and seek deeper meaning in our lives. We seek comfort and meaning in various ways, from religion to science, from shared rituals to impassioned personal pursuits. Some of these pursuits are a form of facing up to our existential anxieties, others a form of escaping them. A colleague once told me why he worked seven days a week, sometimes around the clock, on his research projects: "If I don't work at solving these scientific puzzles, I'll think about death and become riddled with anxiety and depression. I work like this to stave off becoming morose."

We humans spend a lot of energy not facing reality. Our avoidance strategies can take many forms, from workaholism like my scientific colleague's to obsessions with our appearance. Sometimes we become absorbed in the day-to-day realities of meeting our essential needs—because much if not most of the time we do need simply to do our homework, go to work, take out the rubbish, walk the dog, and brush our teeth. We may also seek comfort in the world of the physical, absorbed in the consumption of material goods or getting addicted to the thrill of adrenaline-pumping activities. Yet these are temporary escapes. When we pause from our driven behaviors, we can become overwhelmed with anxiety or lost in a sense of internal emptiness. Without what I am calling temporal integration, we will drift to the banks of either chaos or rigidity.

Our human ingenuity and technical skill can mask our fundamental insecurity. Even the first human being who started a fire using friction and flint must have felt a new command over nature. Knowledge meant survival, whether we found food by distinguishing safe plants from poisonous ones or could predict the seasonal migration of zebra and wildebeests. We have an innate drive to seek out predictable situations. We also come hardwired with a preference for familiar faces—the brain's basic system for knowing whom to trust or discerning who is a member of our clan. These ancient sensations, these drives to feel connected and to be certain, are often directly in conflict with the demands of contemporary culture. We can spend an entire day in a modern city recognizing no one, seeing literally thousands of faces and losing ourselves in anonymity. Our global society, dominated as it is by our drive for mastery, also gives us too much knowledge—flooding us with news of the incalculable daily disasters that can destroy our security in an instant. What happens *there* is known *here* as quickly as the click of a mouse.

What can we do? Our species adapts, learns to make do, to live in megacities of millions, bombarded by information from around the planet. But many of us find that we either numb ourselves to cope or we become painfully aware of the fragility of our condition. How do we find peace of mind? Where are the spaces, the mental sanctuaries, where we can put our heads down on a pillow, certain of our personal and collective survival? The longing for simplicity and shelter still stirs in our synaptic circuitry.

ENTER THE CHECKER

Sandy was twelve years old, and she knew she shouldn't be afraid of the sharp edges of desks, or worried that sharks might be swimming in the next-door neighbor's pool. Behind the long fringe that hung over her face, she looked both embarrassed and terrified as she described these fears and the rituals she had developed to deal with them.

Sandy's parents told me that for the first four months of the school year, she'd done well in her new middle school, had made new friends, and was getting along well with them and with her younger brother. But over the last six weeks, Sandy had become plagued by troubling fears and compulsive behaviors.

Sandy told me that whenever she thought about the edges of desks, or about sharks, she had to count in her head or tap an even number of times with the fingertips of both hands. It turned out that she worried about other disasters as well: earthquakes destroying her house (this was Los Angeles, after all), fires raging through town. She queried me quite intently about the possibility that a shark might find its way into the sewers, come up into the toilet, and bite her. The external facts were that there *had* been a recent quake, the hills just north of town *had* been ablaze, and a surfer in Malibu *had* been attacked by a shark. Seeing these items on the news had provided some of the content of her obsessions, but Sandy's mind was clearly primed for danger.

I asked Sandy what would happen if she didn't tap—or didn't count to an even number in her head. She paused for a moment, looking scared, and then said—"I don't want to find out." She talked more about fires and earthquakes—and her fear of sharks in her neighbors' pool. The weekend before Sandy's family brought her to see me, she had sat at the edge of the pool during an entire afternoon party and wouldn't even put her feet in.

It seemed to me that Sandy might be developing a form of anxiety called obsessive-compulsive disorder (OCD). OCD is characterized by recurrent thoughts—obsessions in the form of frightening images or irrational ideas. Individuals with OCD often feel "stuck" in a thought pattern or behavioral habit they just can't escape. They may have a persistent sense of self-doubt—a "doubt hiccup"—that drives them to check repeatedly to make sure they've locked a door or turned off the stove. They may also exhibit outward behaviors—compulsions such as prolonged, repeated hand washing—that are triggered by their internal sense that something is "just not right." If they enact the compulsion, or think in a certain way, such as counting or repeating a special verbal formula, they believe that bad things will be avoided. And they often worry that if these obsessions or compulsions are not enacted correctly, something will go seriously wrong: Someone might die or become ill, and they would be responsible for failing to prevent it. Still others suffer from a conviction that they are murderers, child abusers, or otherwise immoral people—and that the obsessive-compulsive behaviors will somehow wipe out these crimes or prevent their being enacted in the first place.

OCD can come on suddenly following exposure to the streptococcus bacterium; a protein on its surface elicits an immune response that can irritate the neural circuits that underlie OCD. But Sandy had no history of strep infections, no obvious recent stressors, no accidents, and no big changes in her family life. The only significant event that had occurred was starting middle school—something I made a note to discuss with her later if she came to me for therapy.

Some doctors who diagnose OCD, whatever the situation surrounding its onset, offer antianxiety medication immediately, even in children. But given the potential side effects of these drugs, especially in growing children, and given the fact that they alleviate symptoms only, and only for as long as they are taken, I felt that we should start with a different approach. Research with adults had shown that offering a cognitive and behavioral approach to therapy, combined with mindfulness techniques and information about the brain, could work as well as medication—and with more lasting effectiveness. We didn't have such studies for children, but I had developed a strategy that adapted similar approaches to the developmental needs of the child. My own clinical experience had shown that the same strategies that worked for teens and adults worked for children, too.

Another reason I was open to exploring a medication-free approach was that the onset of Sandy's OCD was recent, and it was not as disabling as some chronic and severe cases I'd seen. If the worry circuits are repeatedly activated over a long period, they can become engrained in the brain and much harder to alter, but since Sandy had been afflicted for a relatively brief period, I felt that I could wait to see if the strategies I proposed to use would work. If the early signs were not encouraging, we could try other cognitive approaches, or, if necessary, turn to medication.

I certainly wanted to give Sandy some immediate relief from her intrusive thoughts and from the ritual behaviors that were taking her over. But I also wanted to offer her the chance to develop new self-regulatory skills—at the level of her brain—that might last a lifetime.

My first goal was to demystify Sandy's condition somewhat so she'd feel less "crazy" and frightened by what was going on in her head. With her parents present, I told her that we each have certain brain circuits that have evolved over millions of years to keep us safe. Using my hand model of the brain, I explained that this circuitry involves the fight-flight-freeze system of the brainstem, the fear-producing amygdala of the limbic area, and the worrying and planning prefrontal cortex. The activation of survival reflexes and the emotion of fear push our cortical areas to find danger—sometimes when a threat is truly there, and sometimes when the sense of danger is only our brain's creation. Because this brain system checks for danger, I like to call it "the checker."

The checker has survived over hundreds of millions of years, I said. It was helping animals long before there were humans, and it takes its job seriously. What would happen, I asked Sandy, if the checker took a long vacation and you were crossing the street? Sandy's eyes widened and she exclaimed: "You'd get run over by a car or a truck!" Exactly. So prehistoric animals without checkers didn't survive—they wouldn't check for saber-toothed tigers at the watering hole, and they'd be eaten up before they ever reproduced. I knew Sandy understood the basics of genetics and evolution when she added, "Yeah—only those animals that had checkers made it, so then their babies had checkers, and they made it, too."

But sometimes, I added, our well-meaning checker gets a bit overexcited. Think of it like this, I said: If a friend comes over and wants to ride a bike with you (Sandy had already told me she liked bike riding) but wanted to pedal forty-five hours nonstop, what would you say? Sandy laughed and said, "No way."

Fine. But instead of just saying no, what if you offered an alternative that you could both live with? What if you said—"Yes, let's ride. But let's ride for just forty-five minutes, not forty-five hours!" Sandy agreed that she and her friend could both have fun that way.

The same is true with the checker, I went on. The idea is to have the checker curb its enthusiasm, and for you to realize that, deep inside, the checker just wants to protect you.

By the end of that first evaluation session I could sense that Sandy felt a bit of relief. She knew that we all have checkers. Some of us have more active checkers than others, but this was a normal part of being human. With this knowledge Sandy and her parents were open to my teaching her some mindfulness practices and other techniques for working directly with the checker. I was not entirely surprised when her mother said that she too struggled with some similar issues and asked if she could join some of my sessions with Sandy. Sandy had never heard about her mother's problem with the checker, and she was happy that they could work on this together. Her mother's openness about having similar symptoms revealed some possible genetic vulnerability to anxiety or OCD, but I knew that we could still do the work necessary to change their brains.

A TIME OF GREAT UNCERTAINTY

During the first stage of Sandy's treatment, I met with her parents, with Sandy alone, and with the entire family so I could hear their different perspectives on what might be going on. In private time with Sandy, I explored whether she'd had any dangerous personal encounters, arguments that left her fearful, or incidents of inappropriate touch. She reported none, and I thought that her onset of symptoms might be understood partly as a result of the change in her school and the sudden changes in her body and feelings as she went through puberty.

As I discussed in chapter 5, the prefrontal cortex is being remodeled during the preteen and adolescent years, and these changes in the brain may themselves be enough to disrupt the ability to self-regulate in the face of fear. Having an overactive checker during this period is not rare. If you look back at your own adolescence, you may remember various rituals and repeating thoughts, coupled with time-honored superstitions (knocking on wood, never walking under ladders, wearing a special shirt on match days) that are relatively mild versions of the checker at work.

If, in addition, Sandy had a genetic predisposition for anxiety, hearing the news of recent natural disasters might have put her fear circuits on high alert. What could she do to feel in control, to soothe herself, when the world around her—and her own adolescent world within—was filled with uncertainty? One way was to behave "as if"—as if she could shape the outcome of events by her own efforts. Enter the checker.

The checker is the neural pinnacle of prediction. There's nothing like the checker's system of dealing with danger to help us, at least on the surface, to take on uncertainty. The checker creates a three-part strategy I like to call SAM. First, the checker Scans for danger, ever vigilant for what might hurt us. Next, the checker sounds an Alert of fear and anxiety whenever something threatening seems to be about to happen. And finally, the checker Motivates us to take action to prevent the danger from occurring.

Under normal conditions, the checker reminds us to look both ways before we cross the street, creates a spike of alarm when we see a truck hurtling towards us, and then motivates us to get out of the way—either by staying put on the sidewalk or speeding up to get to the other side. That's the checker in its most helpful role, and Sandy needed to know that the checker's SAM process was her friend and guardian.

But if the checker gets overactive and works with too much enthusiasm, then we can be paralyzed by its activities. The checker may constantly imagine the worst-case scenario, even when nothing is in fact at risk. The overactive checker adopts the strategy that the best defense is to be prepared for the worst—then you'll never be taken by surprise. When the checker goes overboard, its excesses can take the form not only of hypervigilance and alarm, but also of the obsessive thoughts and compulsive behaviors that are typical of OCD and are irrationally believed to help prevent disaster. Despite the fact that many OCD patients are painfully aware that their behaviors and thought patterns don't make sense, the checker creates an intolerable internal sensation that something needs completion, something must be enacted, and performing the behavior can relieve—if only temporarily—this nagging sense of dread. This, as one of my young patients said, is OCD: Overactive Checker Deployment.

Now imagine that you've listened to the obsessive thought or carried out the compulsive ritual. If nothing bad has happened—no earthquakes, fires, or shark attacks have occurred—your brain has convinced itself that your OCD actions are the reason for your survival. The checker was right: It's a matter of simple cause and effect! The checker has succeeded in its efforts to keep you and others safe, so its strategy is reinforced. Patients are often convinced of this truth with life-or-death intensity. After all, the checker is devoted to our survival—and to passing along our checker genes for another hundred million years—so this is no joking matter.

FOCUSING THE MIND TO CHANGE THE BRAIN

You might be wondering how an intervention that involved more reflection on the internal world could help someone who already was troubled by anxiety and obsessions. Shouldn't Sandy be helped to "get on with life" rather than focusing even more deeply on her mind? In fact, this approach—helping Sandy reframe her symptoms as part of a normal but overactive brain circuit and teaching her mindful awareness strategies—works in two ways. It calms the patient and helps to alleviate symptoms, and it also begins a process of bolstering the self-regulatory circuits in the brain.

At the beginning of our second session, I reviewed the concept of the overactive checker and we discussed how that week at home and school had gone. Then I taught Sandy and her mother the basic meditation I've used throughout this book. They learned quickly to enter a state in which they could sense their breath and become aware—"like I'm watching myself from outside myself," Sandy later told me. She and her mother agreed to practice together each morning for five or ten minutes. Like many children and adolescents, Sandy told me that at times she felt it was "weird" just "sitting watching myself in my head." But soon, as she became more comfortable with the practice, this sense of observation came with a feeling of relief. Sometimes, she realized, she could just "sit" with herself and not have to do anything about her thoughts.

This emerging sense of discernment did not, by itself, eliminate Sandy's worries or her drive to tap, but it did begin to lessen their intensity. She told me how she just counted silently to herself, or hid her hand under her desk at school so no one would see her tapping, but she was still distressed at the notion of what would happen if she didn't count or tap.

During our third and fourth sessions, I began to address Sandy's compulsive behaviors. I wanted to create a little space between her automatic rituals and the impulse that preceded them. I asked Sandy to try spotting the moment when the checker was just speaking up and beginning to get agitated. What was going on inside at that time? Could she detect some inner sensation of fear or worry or dread? As her capacity for mindfulness emerged, I felt she could begin to recognize her obsessions and compulsions as being the result of the checker's activities. This was a variation on the "name it to tame it" strategy I mentioned in chapter 6—a way to calm limbic firing by recruiting the left mode of processing. If she could identify and label the checker at work and recognize that the checker has its own drives and needs, she could begin to differentiate it from the sheer terror she had been experiencing. Knowing that it's a distinct circuit in her brain—not the whole of her—would provide the crucial step in liberating Sandy from the automaticity of her anxiety-driven thoughts and behaviors.

THANK YOU, SAM

Once Sandy could detect her obsessions or compulsive urges arising, we moved to the third phase of treatment. Now she was not only to observe the checker at work, but also to engage it in the kind of internal dialogue I had described in our first meeting. Internal dialogue—sometimes called "self-talk"—is a normal and important part of the moment-to-moment operation of our minds. I simply wanted to harness this internal conversation to help Sandy soothe her distress.

Sandy liked the idea of talking to the checker. It turned out she had already picked up the SAM of Scan, Alert, Motivate, and transformed it into a name: Sam, short for Samantha. I thought this was an encouraging sign—she was befriending this troublesome part of herself. We started to do role plays around different scenarios. Suppose she was having lunch in her neighbor's garden, and the checker got activated. What would it say?

Checker: "Don't get too close to the edge of that pool. They might jump out and grab you."

Sandy (inside her own head): "Thank you, Sam, for your love and concern. I know you want to keep me safe, and I want to be safe, too. But your enthusiasm is too much, and it's not necessary to keep me safe."

At this stage, I told her, you don't have to change your behavior—but the dialogue needs to begin. It's okay if you want to sit as far from the pool as you can, or if you tap or count. Just make sure you talk with the checker first.

This kind of dialogue is in sharp contrast to the internal battle that often takes place before treatment. Sandy's mother told us how she used to criticize herself when her worries came up: "These worries are so dumb—this is ridiculous—just shut up!" or "I can't believe how stupid I am—what an idiot!" If you have a fight with yourself, who can win?

When we see the checker as an alternative state of mind that needs to be embraced, not destroyed, progress can unfold. Why embraced? Because a circuit that has been helping our ancestors survive for millions of years needs to be appreciated for its hard work. If it had failed at its job, you wouldn't be here. Also, whether you're twelve or ninety-two, you're probably not going to win a battle against a brain circuit that's at least one hundred million years old. In an integrative approach, the winning strategy is respect and collaboration.

Sandy's new relationship with the checker gave us an opening for the next step: reducing her ritual behaviors through negotiation. Sandy had been tapping up to fourteen times—always an even number—whenever she experienced her fearful thoughts, and this could happen many times an hour. We had been discussing the motivation behind the tapping ritual—that it was a way of "making sure nothing bad happened." Sandy and I made an agreement for the following week: Whenever the checker told her to tap, she would tap ten times instead of fourteen. Each time, I told her, the checker would probably voice an objection, but Sandy should simply reply, "Thank you for sharing. I know you think tapping will keep us safe, but ten times is just fine." The following week, Sandy would reduce the taps from ten to eight—and in later weeks, down to six, four, and then two. Each time she would continue to thank and reassure the checker.

Of course, I could only hope that nothing happened accidentally while Sandy was decreasing her tapping. Luckily, there were no wildfires in the hills or shark sightings off the beaches to encourage the checker to say "I told you so." And Sandy made steady progress. When necessary, even in the middle of her school day, she would focus on her breath to calm herself or use the special-place imagery we'd also developed together. A problem arose only when Sandy went from the last even number, two, down to just one tap, an odd number. Sam, it appeared, loved symmetry, and moving to one turned out to be a bigger deal than decreasing the total number of taps. This stage took several weeks.

The final stage involved Sandy negotiating with Sam about the frequency of these single taps. First she allowed her checker one tap an hour, then five a day, and so on, until she was down to one. Then one afternoon she arrived for her regular session and told me, "You know, I just realized that I didn't tap yesterday."

THE CIRCUITS OF DOUBT

We never did work out why desks—and not kitchen tables or counters or any other flat rectangular surfaces—were the focus of Sandy's fear. Was she perhaps feeling "cornered" by her challenging new school work? And the sharks? As a scuba diver, I'd been trained to fear sharks, but in Sandy's case all it took was seeing one news report of a shark attack to make her afraid even in her own bathroom. Could these have been mental symbols of the middle school boys who stood in the school yard staring at her new body? I made sure Sandy had time in our sessions to talk about her increased academic load, about boys, and about the whole challenging social scene of middle school. But disabling an out-of-control checker usually takes more than discovering the underlying reason for the fear—and sometimes may not even require that we do so. Years ago, when little was known about the neural circuitry involved, therapists would devote a great deal of time to "getting to the bottom" of the symptoms. As a result they sometimes found themselves chasing one object of fear after another in patients with OCD; one fear faded, only to be replaced by another. Working directly with the circuitry beneath the fear offers a direct route to alleviating it.

The overactive circuits of OCD involve the same areas of our middle prefrontal cortex that alert us when we've made a mistake. In ordinary circumstances, such as my experience with Pablo's antibiotic, a prefrontal area activates the nearby anterior cingulate cortex to create a sense of anxiety. As I discussed in chapter 7, the ACC connects emotion and bodily functions, so the anxiety affects our heart and intestines, giving us an internal sense of dread, which in turn motivates us to find the mistake and correct it.

In OCD, another area deeper in the brain, called the caudate nucleus, is also highly active. The caudate helps us "shift gears" so that we can change the direction of our thoughts or course of action, which is essential to fixing a mistake. But if this prefrontal–caudate link gets stuck in the "on" position, it can create a loop of never-ending worry and agitation. (Strep infections are believed to trigger OCD because they irritate the caudate.) This out-of-control circuit, in turn, can activate the deep alarm system of the brainstem. The brainstem survival reflexes, coupled with the emotion of fear, feed back to the cortical areas, motivating us to search for danger—whether or not it is truly there.

In effect, we were reverse-engineering Sandy's fears. Some alarm stirs in the brainstem and then is picked up and amplified by the fear-generating amygdala. The signal goes to the cortex: "Something is wrong, something is dangerous! Do something!" Now the cortex gets involved and narrows the focus onto a specific item—desk corners, sharks, anything that can give the internal fear state a reason to exist, or that can rationalize why we feel fear in the first place. Next the cortex concocts internal behaviors (obsessive thoughts) or external behaviors (compulsive rituals) to prevent the (imagined) threat from harming us. An integrative mindsight approach recognizes that the checker is trying to keep us safe, to give us some sense of control or certainty in an uncertain world.

A positive stance of collaboration is essential to this work, or else the whole strategy will fall apart. This is one reason mindsight is such a potent tool: It teaches us to be curious, open, and accepting towards whatever arises in our minds. Learning to observe and label, to dialogue and negotiate, Sandy could now monitor her internal world and then modify her thoughts and behaviors. She could have an urge and choose not to turn it into an action.

Sandy's symptoms diminished dramatically within four months, and by six months they were essentially gone. She stopped therapy, although she elected to return for periodic visits, which have been fun for both of us. Now, three years later, Sandy has developed quite a deep wisdom about the nature of her mind, and about being a person on the planet. She no longer worries about going near the edges of swimming pools; she's free to dive right in. She has told me that at times she still hears an intense thought in her head telling her something bad may happen, especially when she's stressed. When this happens, and she feels pushed to start tapping, she engages in some soothing self-talk—"Thank you for worrying about me, Sam, but I can handle this"—and then goes on her way without much difficulty. The checker has been transformed from an oppressive prison guard into a friendly "internal sentry" who watches out for her. This is a resource she'll carry with her, I hope and expect, for the rest of her life.

ACCEPTING UNCERTAINTY

As I made sure Sandy understood, there is nothing inherently wrong with our innate drive to scan for danger, to alert ourselves and others to things that can hurt us, and to do whatever we can to keep ourselves safe. Certainly after the death of Prince Junior, my own checker found a reason to become much more active, and practicing medicine is one continuous lesson in embracing the need to check. Yet experience also teaches us the limits of our control. Even with our best efforts, accidents happen. Life is unpredictable. Temporal integration requires that we let go of the illusion of certainty so that we do what we can to be safe but then release our minds from irrational striving for omniscience and omnipotence.

The beautiful serenity prayer used in Alcoholics Anonymous evokes this letting-go process: "May I have the serenity to accept the things I cannot change, the courage to change the things I can, and the wisdom to know the difference." Serenity, courage, and wisdom are at the heart of temporal integration.

A close friend, Angela, a woman who is like a sister to me, recently developed a rare and life-threatening medical condition. She was taken to a community hospital where her doctor had admission privileges and where she was attended to by a host of specialists. When we spoke on the phone, I asked her if she'd like me to find an academic researcher who might have specialized knowledge about her condition. She said, "Go ahead if that makes you feel better." Of course, I didn't think of it as about my feelings, but about her "proper" care. And in fact I found a researcher who had recently moved to UCLA who specialized in the precise problem she faced. I called Angela back and told her that we could have her transferred to the university hospital for treatment. She refused. She said that she felt comfortable with her current medical team, and that as a recovering alcoholic it was important to her to feel connected to familiar people she already trusted. She thanked me for the consultation and we hung up the phone.

I wondered what to do. Angela sounded rational, but I knew that her condition might be clouding her thinking. However, if she were transferred and the surgery that seemed imminent didn't work, how would I feel? How much should I intrude, even with the intention of saving her life? I called her partner to discuss the benefits of going to a university hospital, and she told me she agreed with Angela—she should decide where she felt comfortable. Then I called Angela back to say I understood her decision and simply to ask her how she was doing. She sounded so strong—filled with the serenity, courage, and the wisdom of her years in AA.

Fortunately, the surgery went well and Angela is doing fine now. But I recognized how strongly the threat of death brought up my drive to control. We want to believe health and youth can belong to us forever; we want to deny the reality of transience in our lives. Sometimes it's good not to accept the first medical solution that's presented to us and to seek out a second opinion that might offer a different diagnosis or treatment plan. But at other times attempts at control are simply an effort to avoid the reality of uncertainty. Serenity, courage, and wisdom: These mindful traits emerge when we acknowledge the mind's drive for certainty and permanence and then refocus our attention on accepting our place in the order of things.

THE COMFORT OF OUR CONNECTIONS

I want to close this chapter by telling you about Tommy, a twelve-year-old patient of mine who became obsessed with death. I had seen him three years earlier, after the death of an uncle with whom he was close. At nine, Tommy was struggling with the first loss in his life, and it changed the way he saw the world. Acknowledging his pain, the fears he'd had about losing his uncle, and then his grief after his uncle died had helped him through the crisis. Over the six-month therapy, and with the help of his parents, he came to feel secure in his family again, and he returned to playing with his friends. During the three years since, his mother told me, he'd been a happy, seemingly carefree kid. But now Tommy had become convinced that he would die from some natural disaster before he turned sixteen. Even when he was not worrying about this calamity, he told me, he was thinking "all the time" about what it would be like to grow old and die.

"Why are we even aware that we die?" he asked, his eyes drilling into mine. I felt his anguish, and Tommy spontaneously brought up his uncle. After an early loss, children often revisit their grief in different ways at each developmental stage. Since he was now entering adolescence, I knew that Tommy's prefrontal changes were allowing him to think about his uncle's death in a larger, more abstract context, and to connect it with his own mortality. I told Tommy how his brain was developing, and that he was now acquiring the prefrontal ability—and the burden it brings—to sense the passage of time and the reality of death. Given these changes in his brain and the new suffering caused by his incessant existential worries, I thought it was time to teach Tommy some mindfulness skills.

He responded well even to our first meditation. He said that he'd "never felt so peaceful, this is incredible! I feel like nothing is wrong, like everything is going to be okay. This is amazing." We continued to practice mindfulness meditation during the next few sessions, and I asked him to practice at home for about ten minutes each morning. I'd introduced him to the image of the ocean and to the peaceful place beneath the surface. I hoped that focusing on his breath would bring him to those tranquil inner depths where he could see his death worries as just brain waves on the surface of his mental sea, so that he could watch them float in and out of his awareness without being so frightened by them. I encouraged Tommy to simply notice his worries, his thoughts, his fears, and not judge them—not try to push them away or banish them from his awareness—but to accept them as just activities of his mind.

Near the end of one session, Tommy told me he had made a discovery. "I realize that if I am known by someone, like my family or my friends, then when I die I won't be gone. Being known makes me feel relaxed. I don't worry."

We sat quietly, reflecting together on that profound insight. His eyes widened and he said, "If I'm known I won't disappear. And when I die I just become a part of everything."

I nodded my head.

"I'll meditate on that," Tommy said.

"I'll meditate on that, too," I said. And then our session ended.

Tommy and I had become fellow travelers on this path of life. As we join with one another, parent and child, patient and therapist, student and teacher, reader and writer, we will find no end to our questions. There is only the ongoing challenge of remaining open to whatever may arise, pain and pleasure, confusion and clarity, step by step along our journey through time.

Epilogue
Widening the Circle: Expanding the Self

IN 1950, ALBERT EINSTEIN RECEIVED A LETTER from a rabbi who had lost one of his two daughters to an accidental death. What wisdom could he offer, the rabbi asked, to help his remaining daughter as she mourned her sister? Here is what Einstein replied:

> A human being is a part of the whole, called by us "Universe," a part limited in time and space. He experiences himself, his thoughts and feelings, as something separated from the rest, a kind of optical delusion of his consciousness. This delusion is a kind of prison for us, restricting us to our personal desires and to affection for a few persons nearest to us. Our task must be to free ourselves from this prison by widening our circle of compassion to embrace all living creatures and the whole of nature in its beauty. Nobody is able to achieve this completely, but the striving for such achievement is in itself a part of the liberation and a foundation for inner security.

BREATHING LIFE ACROSS THE DOMAINS OF INTEGRATION

I initially began to work with people, young and old, within a framework of promoting the eight domains of integration. Coming to therapy with patterns of rigidity or of chaos in their lives, the individual or couple and I would explore how they'd become stuck and identify areas in which differentiation or linkage was blocked. As our work together progressed, as those domains were strengthened and the triangle of well-being was stabilized, I often observed the gradual emergence of a new sense of wholeness. In various ways, my patients expressed their growing awareness of being a part of a larger whole, of inhabiting a bigger world than before they started the work. In direct opposition to the common concern that psychotherapy and contemplation are "self-indulgent" activities, clarifying mindsight's lens actually led to a very different state than self-involved navel-gazing or inward preoccupation. Instead it seemed that the mindsight work led directly to a deep sense of wanting to give back to others, to expanding the focus of concern, and to the identification of a larger set of causes. This sense of being a part of a whole beyond their immediate relationships and social worlds led directly to the wider "circle of compassion" of which Einstein wrote.

This happened in small ways, as when twelve-year-old Tommy wrestled with his worries about death. As we worked together on what I've called "temporal integration," Tommy discovered for himself that feeling connected to others lessened his feelings of dread and isolation, so that "when I die I just become a part of everything." It happened in larger ways with Matthew, the lonely man with the revolving door of frustrating romances. As he worked through his states of shame and isolation, Matthew came to feel a desire to get involved with something beyond himself, and he found that working to heal the bay that borders our community filled him with purpose and passion. He could mobilize his business skills and connections to preserve our natural resources, not only in ways that benefited people alive now whom he didn't know, but to support future generations. Others, such as Peter and Denise, who became advocates for each other through interpersonal integration, were moved to donate to charities that provided support to families devastated by illness. Certainly this had meaning for them personally, given their tragic childhood losses, but their immediate circle of concern gave way to a broader sense of commitment.

I didn't then have a simple way to describe this observation—that as people became more integrated in the first eight domains, their sense of identity expanded. The boundaries of "self" became wide open. I chose the term *transpiration* to connote the way we "trans" (*across*) "spire" (*breathe*), how we breathe across the eight domains of integration. Transpiration is how we dissolve our sometimes confining sense of an "I" and become a part of an expanded identity, a "we" larger than even our interpersonal relationships. It is how we "integrate integration."

It's important to note that achieving this cross-breathing way of being did not require anything special beyond developing the basic reflective skills of mindsight. The people who arrived at transpirational integration started from very different places and were motivated to develop the domains of integration for many different reasons. Some were faced with an immediate challenge, a conflict urgently needing resolution. Others were dealing with the pain of an unresolved loss from long ago, an unhealed trauma or defeat. There was no singular path that seemed necessary beyond coming to a deeper, reflective sense of the mind, of seeing the inner world with more acuity and promoting integration across these various domains.

For as long as we have had records of them, contemplative practices have described a similar sense of the true interconnectedness of all things. But for much of our history as a species—and perhaps particularly in modern society—we have often seen ourselves as isolated beings, solo actors on a small stage with a few select fellow thespians. Other theaters are not important, and perhaps even competitive with our performance. Why would we be so confined in how we define ourselves?

ME VERSUS THEM

Today we can actually track scientifically the neural dimensions of our narrow definitions of self. When our resonance circuits are engaged, we can feel another's feelings and create a cortical imprint that lets us understand what may be going on in the other's mind—because it is like ours—and our mind and our brain turn on our mindsight mechanism. We uncap our inner lens and take a deep look into the face of the other to see the mind that rests beneath the visage. But if we cannot identify with someone else, those resonance circuits shut off. We see others as objects, as "them" rather than "us." We literally do not activate the very circuits we need in order to see another person as having an internal mental life.

This shutting off of circuits of compassion may be one explanation for our violent history as a species. Without mindsight, people become objects, rather than subjects themselves with minds like ours worthy of respecting and even knowing. Under threat, we may distort what we see in others, project our own fears onto their intentions, and imagine that they will harm us. We may also perceive malevolence where none exists, and then retreat to the fight-flight-freeze survival reactions of a threatened state of mind. If the threatened state creates within us a "fight" response, then we get the object out of our way however possible.

Feeling threatened takes over our perception. Sometimes this is to our benefit, as when I "saw" that snake just steps ahead of my son on a mountain trail and later became aware of feeling fear. But at other times the same brain mechanism can dramatically affect the way we behave towards others. Imaging studies have demonstrated that when we are shown photographs suggesting danger and threat, such as a gun pointing at us or a close-up of a fatal car accident, our brains go on high alert. Even when the images are shown so rapidly that we cannot detect them consciously, these subliminal displays affect our mental states and our behavior. Such "mortality salience" studies have shown repeatedly that with people "like us," we become kinder and extend ourselves more to care for their welfare. They are seen as members of our clan, fellow inhabitants of our cave, and we protect them from the harm we've been primed subliminally to fear. If, on the other hand, the people are "not like us," we are more likely to treat them with disdain and disregard—as if they were potential enemies and perpetrators of harm. We banish them more easily, create more intense punishments for any wrongdoing, and judge them more harshly.

Without awareness of these mechanisms of the mind that classify "like me" and "not like me" during moments of threat, our humanity is at risk. In our global, instant-information, high-tech world, not having the mindsight to disengage these rapid, subcortically driven alarms can have dire consequences.

When we become survival-driven, we lose any or all of the nine middle prefrontal functions that Barbara lost when she "lost her soul," and we are primed to travel down a low road. When we are reactive, we revert to primitive behaviors without flexibility or compassion. We act impulsively, lose the ability to balance our emotions, and fail to exert moral reasoning. Both individual behavior and public policy can be shaped by these unexamined autopilot neural responses. Instead of being guided by understanding and compassionate concern, even for those who threaten us, our mind-sightless response is to become hostile and inflexible, and to lose our moral compass.

EXPANDING IDENTITY

The study of positive psychology suggests that being involved in something larger than a personal self creates a sense of meaning and well-being—an essential part of the experience of "happiness." When we spend money on others, for example, we feel more content than when we spend money on ourselves. This is a kind of well-being rooted in meaning, connection, and equanimity—called *eudaimonia* by the ancient Greeks and in modern times perhaps called "inner" or "true" happiness. Ironically, being personally happy requires that we greatly expand our narrowly defined individual pre-occupations. We are built to be a "we"—and enter a more fulfilling state, perhaps a more natural way of being, when we connect in meaningful ways with others. A living organism links its differentiated parts—and without this integration, it suffers and dies.

Science has shown that well-being and true happiness come from defining our "selves" as part of an interconnected whole—connecting with others and with ourselves in authentic ways that break down the isolative boundaries of a separate self. Such connections can be created through developing the clear lens of mindsight, which enables us to track energy and information flow within and among us. Cultivating our capacity to sense energy and information flow helps us expand the "self" beyond the boundaries of our body and reveals the fundamental truth that we are indeed a part of an interconnected world. Our "living organism" is the extended community of living beings.

This proposal is no easy task. Dissolving fixed mental perceptions created along the brain's firing patterns and reinforced relationally within our cultural practices is no simple accomplishment. Our relationships engrain our early perceptual patterns, deepening the ways we come to see the world and believe our inner narrative. Without an internal education that teaches us to pause and reflect, we may tend to live on automatic and succumb to these cultural and cortical influences that push us towards isolation.

We need to examine directly the ways in which our cortical processes create the top-down influences from prior experience that cloud our vision. Part of our challenge in achieving well-being, in ourselves and perhaps in our world, is to develop enough mindsight to clear us of these restrictive definitions of ourselves so that we can grow towards higher degrees of integration within our individual and collective lives.

SEEING CLEARLY

If the mind creates automatic constraints on our sense of self so that we tend to see ourselves as separate from one another, how do we take the steps as individuals and as a society to widen our circles of compassion and dissolve these automatic top-down processes? The effective strategy seems to be to help one another see the mind clearly.

Seeing the mind clearly not only catalyzes the various dimensions of integration as it promotes physical, psychological, and interpersonal well-being, it also helps us dissolve the optical delusion of our separateness. We develop more compassion for ourselves and our loved ones, but we also widen our circle of compassion to include other aspects of the world beyond our immediate concerns. This transpirational awareness gives us a sense of being a fundamental part of a larger world. Physical separations and differences become less paramount as we see that our actions have an impact on the interconnected network of living creatures within which we are just a part. Time separations and distances also become less self-defining as we see ourselves as a fundamental link between what came before and what will exist long after these bodies are gone from this life. This is the essence of transpiration.

With integration, we see ourselves with an expanded identity. When we embrace the reality of this interconnection, being considerate and concerned with the larger world becomes a fundamental shift in our way of living. When we sense the importance of our caring for one another and the planet, we can see that beyond creating meaning and happiness, transpiration and the integration from which it grows may be essential for our survival.

Physically and genetically, our brains may not have evolved much in the last forty thousand years—but our minds have. A baby born today would be much the same as a baby born tens of thousands of years ago. But if we were able to compare the intricate neural structure of an adult brain in today's modern society with that of an adult brain from forty thousand years ago, we'd find huge differences. With markedly contrasting culturally shaped experiences, the mature brain in each environment would have responded to the energy and information flow with strikingly different neural connections.

The mind uses the brain to create itself. As patterns of energy and information flow are passed among people within a culture and across generations, it is the mind that is shaping brain growth within our evolving human societies. The good news about this perspective from science is that we can use an intentional attitude in our modern lives to actually change the course of cultural evolution in a positive direction. Cultivating mindsight in ourselves and in one another, we can nurture this inner knowing in our children and make it a way of being in the world. We can choose to advance the nature of the mind for the benefit of each of us now and for future generations who will walk this earth, breathe this air, and live this life we call being human.

Acknowledgments

This project has been developing over my entire life, and many, many people have shared the journey leading to the ideas for this book. From the first days of medical school, my patients have served as a primary motivation for forming the principles underlying mindsight. It has been a profound privilege to be invited into their lives, to share the pain and confusion of their struggles, the clarity and joy of their triumphs. Through the opportunity to be a part of their efforts to understand and change their lives—to travel with them on an expedition to transform anguish and despair into resilience and empowerment—I have been given insights both personally and professionally in ways I could never have dreamt possible. These insights are the direct source of motivation to write this book—and it is the collected wisdom of my patients' struggles that has given voice to the notion of mindsight.

I am deeply appreciative of my teachers for their guidance during my clinical and research training in psychiatry: Drs. Gene Beresin, Leston Havens, David Herzog, and the late Tom Whitfield in medical school; Drs. Gordon Strauss, Joel Yager, and the late Denny Cantwell and Robert Stoller in residency; and Drs. Robert Bjork, Chris Heinicke, Eric Hesse, Mary Main, and Marion Sigman in research training.

I thank my clinical colleagues for their support over the thirty years since I began my medical training. Students and colleagues at the Mindsight Institute have also been a wonderful source of lively discussions as the field of interpersonal neurobiology has been emerging. Many read earlier versions of this manuscript and I thank them for their suggestions and questions over the years. Especially helpful have been Bonnie Badenoch, Eric Bergemann, Tina and

Scott Bryson, Lynn Cutler, Erica Ellis, Donna Emmanuel, Stephanie Hamilton, Joan Rosenberg, and Aubrey Siegel. I also thank my interns, Gabe Eckhouse, Deanie Eichenstein, and Ellen Streit, together with Beth Pearson, Tom Pitoniak, and Kate Norris at Random House, for their insightful immersion in the manuscript's final editing. Brian McLendon and Carolyn Schwartz have also been an important part of Random House's efforts to bring this book to the public eye.

The Global Association for Interpersonal Neurobiology Studies has been a great gathering of like-minded people who share this notion that seeing the mind can enhance our individual and collective lives. I first began my career in the academic world and am happy as a clinician to keep one foot planted in scientific pursuits at the Foundation for Psychocultural Research/UCLA Center for Culture, Brain, and Development (CBD) and at the Mindful Awareness Research Center (MARC). The individuals associated with those two university centers provide the intellectual grounding that a synthesizer working in the subjective world of psychotherapy needs to continue to be challenged and stimulated to think rigorously. I thank Sue Smalley, Diana Winston, and Susan Kaiser Greenland at MARC; and Mirella Dapretto, Patricia Greenfield, Eli Ochs, Alan Fiske, Marco Iacoboni, and Allan Schumann at CBD. I also want to thank all the writers who have contributed to the interpersonal neurobiology series, where we try to bridge research, clinical practice, and education—especially Lou Cozolino and Allan Schore, who have also been friends and colleagues since the first days of this effort to synthesize science and psychotherapy. At the Lifespan Learning Institute, I am grateful to Marion Solomon and Bonnie Goldstein for their never-ending support and camaraderie. I'd also like to express my gratitude to the Atlas, the Attias, and the Kirlin Family Foundations for their support.

Colleagues and friends have been instrumental in my personal and intellectual growth and have furnished invaluable reflections on this work as it unfolded. Diane Ackerman, Dan Goleman, Jon Kabat-Zinn, Jack Kornfield, Regina Pally, and Rich Simon have become siblings in my heart, and I thank them for their friendship and support at many a challenging moment. I also want to remember my dear friend and colleague John O'Donohue, who left this world too soon but whose magnificent books and love of life continue to inspire me every day.

My two children have given me invaluable support; their good humor and lively debates keep me on my toes and teach me never to take anything for granted. As those with adolescents may know, it is not possible to become complacent with who you are when confronted with teenage honesty.

My wife, Caroline Welch, is a fountain of wisdom and encouragement and an inspiration in my life. Caroline read every version of the manuscript as it evolved, and her contributions have been an essential part of the book's creation. I am profoundly grateful for our relationship.

I'd like to give a deep bow of thanks to my literary agent and friend, Doug Abrams, who shares a dedication to bringing ideas into the world that, one page at a time, may move it even one small step in a positive direction. My own life has been enhanced because of our connection. When we set out on this project, we were looking for a publisher who might share its vision. Finding Toni Burbank at Bantam was a dream come true. Over the years since we first met, Toni has kept up her reputation as a brilliant and supportive award-winning publisher. I have been fortunate that she was not only interested in acquiring this project, she also wanted to be the hands-on editor for it. Our relationship grew as we pored over the pages of each chapter, and I've come to learn firsthand that she is not only keenly intelligent and a wizard with words—she is also fun and funny. I appreciate her dedication to this project, and have marveled at her capacity to use mindsight to keep the reader's experience in focus as we wrestled with the details of science and the flow of the narratives. Joining this project as the anchor editor, Beth Rashbaum has also been extremely helpful and a great asset to our team. Her enthusiasm, insightful suggestions, and broad perspective enabled the addition of many vital final touches. I cannot thank Beth, Toni, and Doug enough for being companions on this journey to bring *Mindsight* to life.

Appendix

Here are a dozen basic concepts and related terms and ideas that form a foundation for our approach of mindsight, integration, and well-being.

1. The *Triangle of Well-Being* reveals three aspects of our lives. *Relationships, Mind,* and *Brain* form the three mutually influencing points of the Triangle of Well-Being. *Relationships* are how energy and information is shared as we connect and communicate with one another. *Brain* refers to the physical mechanism through which this energy and information flows. *Mind* is a process that regulates the flow of energy and information. Rather than dividing our lives into three separate parts, the Triangle actually represents three dimensions of one system of energy and information flow.

2. *Mindsight* is a process that enables us to monitor and modify the flow of energy and information within the Triangle of Well-Being. The *monitoring* aspect of mindsight involves sensing this flow within ourselves—perceiving it in our own nervous systems, which we are calling Brain—and within others through our Relationships, which involve the sharing of energy and information flow through various means of communication. We then can *modify* this flow through awareness and intention, fundamental aspects of our mind, directly shaping the paths that energy and information flow take in our lives.

3. A *system* comprises individual parts that interact with one another. For our human systems, these interactions often involve the *flow of energy and information*. Energy is the physical property enabling us to do something; information is the representation of something other than itself. Words and ideas are examples of units of information we use to communicate with one another. Our relationships involve our connection to other people in pairs, families, groups, schools, communities, and societies.

4. We can define *well-being* as occurring when a system is integrated. *Integration involves the linkage of differentiated parts of a system.* The differentiation of components enables parts to become individuated, attaining specialized functions and retaining their sovereignty to some degree. The linkage of parts involves the functional connection of the differentiated components to one another. Promoting integration involves cultivating both differentiation and linkage. Mindsight can be used to intentionally create integration in our lives.

5. When a system is open to outside influences and capable of becoming chaotic, it is called a dynamic, nonlinear, complex system. When this type of system is integrated, it moves in a way that is the most flexible and adaptive. We can remember the characteristics of an integrated flow of the system with the acronym *FACES: Flexible, Adaptive, Coherent, Energized, and Stable.*

6. The *River of Integration* refers to the movement of a system in which the integrated FACES flow is the central channel and has the quality of harmony. On either side of the River's flow are two banks—chaos and rigidity. We can detect when a system is not integrated, when it is not in a state of harmony and well-being, by its chaotic or rigid characteristics. Recurrent explosions of rage or terror and being taken over by a sense of paralysis or emptiness in life are examples of these chaotic and rigid states outside the River of Integration.

7. In this model, eight *Domains of Integration* can be harnessed to promote well-being. These include *consciousness, horizontal, vertical, memory, narrative, state, interpersonal,* and *temporal integration.* As the mind is an embodied and relational process that regulates the flow of energy and information, we can use the intentional focus of our awareness to direct this flow towards integration in both Brain and Relationships. As these domains of integration are cultivated, a ninth domain, *transpirational integration,* may begin to emerge in which we come to feel that we are a part of a much larger, interconnected whole.

8. Integration in relationships involves the attuned communication among people who are honored for their differences and then linked together to become a "we." Integration in the brain—what we are using as a term for the extended nervous system distributed throughout the entire body—involves the linkage of separate, differentiated neural areas and their specialized functions to one another. The focus of our attention directs the flow of energy and information through particular neural circuits. In this way we can say that *the mind uses the brain to create itself.* Attention activates specific neural pathways and lays the foundation for changing the connections among those firing neurons by way of a fundamental process called *neuroplasticity.* The function of our mind—the regulation of energy and information flow—can actually change the structure of the brain itself. Mindsight enables us to create neural integration.

9. One example of neural integration is revealed in the functions that emerge from a highly integrative area of the brain called the *middle prefrontal cortex.* Involving specific parts of the prefrontal region located behind the forehead (including the anterior cingulate, orbitofrontal, and the medial and ventrolateral prefrontal zones), the middle prefrontal integrative fibers link the whole cortex, limbic area, brainstem, body proper, and even social systems to one another. The *nine middle prefrontal functions* emerging from this multidimensional neural integration include: 1) body regulation, 2) attuned communication, 3) emotional balance, 4) fear modulation, 5) response flexibility, 6) insight, 7) empathy, 8) morality, and 9) intuition. These functions would top many people's list of a description of well-being. They are also the established outcome and process of the reflective skills of looking inward, and the first eight of this list are proven outcomes of secure parent-child relationships that are filled with love. This list exemplifies how integration promotes well-being.

10. Mindsight doesn't just emanate from the middle prefrontal cortex. The reflective practice of focusing internal attention on the mind itself with *openness, observation, and objectivity*—the essentials of a strengthened *mindsight lens*—likely promotes the growth of these integrative middle prefrontal fibers. We use the acronym *SNAG* to denote how we *Stimulate Neuronal Activation* and *Growth.* This is the foundation of neuroplasticity, of how experiences—including the focus of our attention—transform brain structure. Mindsight SNAGs the brain towards integration, making it possible to intentionally promote linkage and differentiation within the various domains of integration.

11. A *Window of Tolerance* refers to the band of tolerable levels of arousal in which we can attain and remain in an integrated FACES flow and live with harmony. Widened Windows create resilience in our lives. If a Window is narrowed, then it becomes more likely for energy and information flow to move outside its boundaries and for our lives to become chaotic or rigid. The integrated states within the Window of Tolerance are our subjective experience of living with a sense of ease and in the harmonious FACES flow down the River of Integration. As we *SIFT* the mind—tracking the *Sensations, Images, Feelings,* and *Thoughts* that dominate our internal world— we can *monitor* energy and information flow moment by moment within our Windows of Tolerance and *modify* our internal state to remain integrated and in a FACES flow. Ultimately we can use this monitoring and modifying to change not only our present *state*, but also our long-term *traits* that reveal how our Windows for various feelings or situations can be widened through changes in our brain's dynamic regulatory circuits.

12. The *Wheel of Awareness* is a visual metaphor of the mind. We can stay within the open, receptive *hub* of the Wheel to sense any mental activities emerging from the *rim* without becoming swept up by them. A strengthened hub permits us to widen our Windows of Tolerance as we become more observant, objective, and open and thus attain more resilience in our lives. Mindsight harnesses this important capacity to remain receptive and to be able to monitor the internal world with more clarity and depth. We are then in a position to modify our inner and interpersonal world as we cultivate integration and move our lives towards more compassion, well-being, and health.

Notes

I have served as the founding series editor for more than a dozen textbooks in the Norton Professional Series on Interpersonal Neurobiology (IPNB) that provide extensive scientific references and discussions of practical applications of this exciting new field. These texts contain literally thousands of scientific references regarding this view of mind, brain, and relationships and other topics that are relevant to *Mindsight*. The field of IPNB is first introduced in Daniel J. Siegel, *The Developing Mind* (New York: Guilford, 1999), and its applications to parents are explored in Daniel J. Siegel and Mary Hartzell, *Parenting from the Inside Out* (New York: Tarcher/Putnam, 2003). The books presently in the IPNB series include: *The Neuroscience of Psychotherapy* (Louis Cozolino, 2002), *Healing Trauma* (ed. Marion Solomon and Daniel J. Siegel, 2003), *Affect Dysregulation and Disorders of the Self* and *Affect Regulation and the Repair of the Self* (Allan N. Schore, 2003), *The Present Moment in Psychotherapy and Everyday Life* (Daniel N. Stern, 2004), *The Neuroscience of Social Relationships* (Louis Cozolino, 2005), *Trauma and the Body* (Pat Ogden, Yekuni Minton, and Clare Pain, 2006), *The Haunted Self* (Onno van der Hart, Ellert S. Nijenhuis, and Kathy Steele, 2007), *The Mindful Brain* (Daniel J. Siegel, 2007), *The Neurobehavioral and Social Emotional Development of Infants and Children* (Ed Tronick, 2008), *Being a Brain-Wise Therapist* (Bonnie Badenoch, 2008), *The Healthy Aging Brain* (Louis Cozolino, 2008), *Early Intervention, and Relationship-Based Therapies: A Neurorelational Framework for Interdisciplinary Practice* (Connie Lillas and Janeice Turnbull, 2009), *The Healing Power of Emotion* (ed. Diana Fosha, Daniel J. Siegel, and Marion Solomon, 2009), *A Glossary of Affect Regulation* (ed. Allan and Judith Schore, forthcoming), and *The Mindful Therapist: A Clinician's Guide to Mindsight and Neural Integration* (Daniel J. Siegel, forthcoming).

Further information can also be obtained from the Global Association for Interpersonal Neurobiology Studies (GAINS), which can be contacted at Mindgains.org and through the numerous educational programs found on the MindsightInstitute.com website.

INTRODUCTION: DIVING INTO THE SEA INSIDE

xii **having social and emotional intelligence:** See Daniel Goleman, *Emotional Intelligence* (New York: Bantam, 1994), and his discussion of the related topic in *Social Intelligence* (New York: Bantam, 2008). The capacity for mindsight can be seen as the basis of these forms of inter- and intrapersonal intelligence.

xiii **But even if such early support was lacking:** The concept of *mindsight* and its clinical implications now have empirical support from the studies of similar processes denoted by scientific terms such as *theory of mind, mentalese, mind reading, psychological-mindedness, mind-mindedness, reflective function,* and *mentalization.* A fascinating review of some of these scientific studies is given in *Understanding Other Minds,* Bertram Malle and Sara Hodges, eds. (New York: Guilford, 2005). Also of note is the work of Jon Allen, Peter Fonagy, and Allan Bateman, who have explored mentalization both in attachment and in personality disorders. They have empirically demonstrated that mentalization is absent in impairments to secure attachment and that it is capable of being taught in adulthood. See Jon G. Allen, Peter Fonagy, and Allan W. Bateman, *Mentalizing in Clinical Practice* (Arlington, Va.: APPI, 2008).

xiii **I coined the term mindsight:** I first used this term in print in *The Developing Mind.*

xiii **What has been called our sixth sense:** Though some popular uses of the term *sixth sense* refer to the ability to see the dead (as in the movie by this name) or other perceptual claims, in the 1800s, Charles Bell and later William James apparently used it to refer to our ability to perceive the internal state of the body. (See Pat Ogden, Kekuni Minton, and Claire Pain, *Trauma and the Body* (New York: Norton, 2007). Steve Porges more recently referred to this sixth sense in infants in the *Zero to Three* newsletter October/November, 1993. The "sixth" sense as a term makes sense as we bring perception inward (the first five bringing the outside world into view). This sixth sense would include balance and proprioception—knowing your position in space—as well as the sense of hunger and thirst and internal signals from muscles, teeth, and pain sensors in the skin. Even sensual touch is a part of this interior data. Having a visceral sense—the feelings of your viscera, such as the heart, lungs, and intestines—would also be included here and has been called "entero-ception." Taken together, knowing the internal world can be called "interoception." The spinal cord layer called lamina 1 carries this internal data upward to the various parts of the brain in the skull. We then are in the position to name mindsight our "seventh sense."

xiv *How we focus our attention:* This statement comes from the exciting new research on neuroplasticity. Two especially accessible resources are Sharon Begley, *Train your Mind, Change your Brain* (New York: Ballantine, 2007), and Norman Doidge, *The Brain That Changes Itself* (New York: Penguin, 2007).

xv *achieve and maintain* **integration:** See *The Developing Mind* and *The Mindful Brain* for detailed discussions of integration.

xvi *Scientific studies support this idea:* In the last decade and a half, a number of formal scientific studies have emerged examining how the way we look inward, or "reflect," can have significant impacts on our well-being. Emotional awareness is one such concept, as explored by Daniel Goleman in *Emotional Intelligence*. See also Allen, Fonagy, and Bateman, *Mentalization in Clinical Practice*. Jean Decety and Yoshiya Moriguichi provide an insightful and comprehensive discussion of the ways in which the empathy portion of mindsight is impaired in schizophrenia, borderline and narcissistic personality disorders, antisocial personality, autism spectrum disorders, and the general condition of alexithymia (people not knowing their feelings). See Jean Decety and Yoshiya Moriguchi, "The Empathic Brain and Its Dysfunction in Psychiatric Populations: Implications for Intervention Across Different Clinical Conditions," *Biopsychosocial Medicine* (2007): 1:22. Published online in 2007.

xvi *Research has also clearly shown:* Please see the work of the Consortium for Academic, Social, and Emotional Learning (Casel.org). Mark Greenberg outlines in various texts the reflective skills at the heart of how SEL promotes "executive function" development—a form of cognitive resource allocation that relies on prefrontal function. This work is applied in the teacher's educational program of the Garrison Institute, called CARE: Curriculum for Awareness and Resilience in Education.

xvii *those with autism and related neurological conditions:* It is crucial to keep in mind that mindsight may be impaired in inherited disorders—as well as blocked in its development with suboptimal experiences. Decety and Moriguchi's "The Empathic Brain" offers an overview of this issue. Related work is summarized in Simon Baron Cohen, *Mindblindness: An Essay on Autism and Theory of Mind* (Cambridge, Mass.: MIT Press, 1997), and Mirella Dapretto et al.,

"Neural Mechanisms of Empathy in Humans: A Relay from Neural Systems for Imitation to Limbic Areas," *Proceedings of the National Academy of Sciences* 100, no. 9 (2003): 5497–5502. Marco Iacoboni summarizes his work in *Mirroring People* (New York: Farrar, Straus, and Giroux, 2008) and explores this area of mirror neurons and autism. See also the following explorations of autism and its potential correlates: Justin Williams et al., "Imitation, Mirror Neurons and Autism," *Neuroscience and Biobehavioral Review* 25 (2001): 287–95; Uta Frith, *Autism: Explaining the Enigma* (New York: Blackwell, 2003); Uta Frith and Christopher D. Frith, "Development and Neurophysiology of Mentalizing," *Philosophical Transactions of the Royal Society,* Series B: *Biological Sciences* 358 (2003): 459–73; Simon Baron-Cohen, "Theory of Mind and Autism: A Fifteen-Year Review," in *Understanding Other Minds: Perspectives from Developmental Neuroscience,* ed. Simon Baron-Cohen, Helen Tager-Flusberg, and Donald Cohen (New York: Oxford University Press, 1994); Ami Klin, Robert Schultz, and Donald Cohen, "Theory of Mind in Action: Developmental Perspectives on Social Neuroscience," in ibid.

xvii **Neuroscientists are now identifying:** See the IPNB publications by Schore (2003), Cozolino (2005), Tronick (2008), and Lillas and Turnbull (2009); and see Siegel (1999, *The Developing Mind*).

xvii **If parents are unresponsive, distant:** See the summary of this work in L. Alan Sroufe, Byron Egeland, Elizabeth A. Carlson, and W. Andrew Collins, *The Development of the Person* (New York: Guilford, 2005); Siegel, *The Developing Mind*.

xvii **The good news is that:** See the IPNB publications by Schore (2003) and Tronick (2008); and see Siegel, *The Developing Mind*.

xvii **Here we see living evidence:** The extensive literature on neuroplasticity is summarized in Eric R. Kandel, *In Search of Memory: The Emergence of a New Science of Mind* (New York: Norton, 2007). The concepts of neuroplasticity are covered in Kandel and are reviewed in accessible ways in Sharon Begley, *Train Your Mind, Change Your Brain* (in paperback as *The Plastic Mind*) (New York: Random House, 2007); Norman Doidge, *The Brain That Changes Itself;* and Sandra Blakeslee and Matthew Blakeslee, *The Body Has a Mind of Its Own* (New York: Random House, 2007). For a review of related areas of neuroscience, see the basic text *Principles of Neural Science,* 4th ed., revd., Eric R. Kandel, James H. Schwartz, Thomas M. Jessell, eds. (New York: McGraw-Hill, 2000). Note here, too, that *neuroscience* and *neural science* are synonymous, as are *neuroplasticity* and *neural plasticity.*

CHAPTER 1: A BROKEN BRAIN, A LOST SOUL

5 ***After an injury, the brain can regain:*** See Kandel, *In Search of Memory;* Doidge, *The Brain That Changes Itself;* Begley, *Train Your Mind, Change Your Brain.*

5 **Neuroplasticity** *is the term used:* Kandel, Schwartz, and Jessel, eds., *Principles of Neural Science;* Begley, *Train Your Mind, Change Your Brain;* Doidge, *The Brain That Changes Itself.*

7 ***It also links widely separated:*** For an extensive discussion of the research behind the areas of the middle prefrontal cortex, see Appendix IIIC in Siegel, *The Mindful Brain.*

8 ***I was puzzled by that disconnect:*** See Stanley B. Klein, "The Cognitive Neuroscience of Knowing One's Self," Michael S. Gazzaniga, ed., *The Cognitive Neurosciences,* 3rd ed. (Cambridge, Mass.: MIT Press, 2004); Decety and Moriguichi, "The Empathic Brain"; and Bernard Beitman and Jyotsna Nair, eds., *Self-Awareness Deficits in Psychiatric Patients* (New York: Norton, 2004). For further elaboration on the nature of self-knowing awareness, please see Sterling C. Johnson et al., "Neural Correlates of Self-Reflection," *Brain* 125 (2002): 1808–14. Also see Troels W. Kjaer, Markus Nowak, and Hans C. Lou, "Reflective Self-Awareness and Conscious States: PET Evidence for a Common Midline Parietofrontal Core," *NeuroImage* 17 (2002): 1080–86; Kai Vogeley and Gereon Fink, "Neural Correlates of First-Person Perspective," *Trends in Cognitive Sciences* 7 (2003): 38–42.

8 ***In the years since I took Barbara's:*** Please see Siegel, *The Mindful Brain,* for extensive discussions of the prefrontal region and its functions. Also see Antonio R. Damasio, *Descartes' Error: Emotion, Reason, and the Human Brain* (New York: Avon Books, 1994), which explores the case of Phineas Gage in the 1800s and his accidental injury to this region of the brain. Further discussion of the role of this area of the brain can be seen in Kevin S. LaBar et al., "Dynamic Perception of Facial Affect and Identity in the Human Brain," *Cerebral Cortex* 13 (2003): 1023–33; Andrea D. Rowe et al., "'Theory of Mind' Impairments and Their Relationship to Executive Functioning Following Frontal Lobe Excisions," *Brain* 124 (2001): 600–16; and Simone G. Shamay-Tsoory et al., "Characterisation of Empathy Deficits Following Prefrontal Brain Damage: The Role of the Right Ventromedial Prefrontal Cortex," *Journal of Cognitive Neuroscience* 15 (2003): 324–37.

10 ***Called the "still-face" experiment:*** See the groundbreaking work of Ed Tronick. The most recent compilation of his important

contributions is in the IPNB Series book, *The Neurobehavioral and Social Emotional Development of Infants and Children* (2008).

14 **Minding the Brain: The Brain in the Palm of Your Hand:** The "brain in the palm of your hand" was first introduced in *The Developing Mind* and then first illustrated in *Parenting from the Inside Out*. Elaborated here, these are the basics of neuroanatomy and function that you can explore further in a wide array of textbooks and illustrated atlases on the brain such as Kandel, Schwartz, and Jessel, eds., *Principles of Neural Science;* V. S. Ramachandran, *Encyclopedia of the Human Brain* (San Diego: Academic Press, 2002); and Gerald Edelman and Jean-Pierre Changeux, *The Brain* (New York: Transaction, 2001). For an application of neuroscience in the workplace, see David Rock, *Your Brain at Work* (New York: Harper Business, 2009).

15 **This could be called "horizontal":** Integration in general is highlighted in Siegel, *The Developing Mind.* These "domains" of integration are discussed in Siegel, *The Mindful Brain,* and will be detailed in part 2. For a review of the science of laterality, see Richard J. Davidson and Kenneth Hugdahl, *Brain Asymmetry* (Cambridge, Mass.: MIT Press, 1996).

17 **But whichever of these responses is chosen:** A synthesis of the reactive versus receptive modes can be found in Steven Porges, "Reciprocal Influences Between Body and Brain in the Perception and Expression of Affect: A Polyvagal Perspective," Fosha, Siegel, and Solomon, eds., *The Healing Power of Emotion.* See also Steven W. Porges, "Love: An Emergent Property of the Mammalian Autonomic Nervous System," *Psychoneuroendocrinology* 23, no. 8 (1998): 837–61.

17 **The brainstem is also a fundamental:** See Jaak Panksapp, *Affective Neuroscience* (New York: Oxford University Press, 1998), and "Brain Emotional Systems and Qualities of Mental Life: From Animal Models of Affect to Implications for Psychotherapeutics," in Fosha, Siegel, and Solomon, eds., *The Healing Power of Emotions.*

CHAPTER 2: CREPES OF WRATH

26 **Let me briefly map my meltdown:** Again, any references to the middle prefrontal cortex and its functions can be explored in more depth in Siegel, *The Mindful Brain,* especially in Appendix IIIC. The "middle prefrontal cortex" includes the anterior cingulate,

orbitofrontal, and medial and ventrolateral prefrontal cortical areas. The anterior portion of the insula can be considered to be a part of the ventrolateral prefrontal region.

29 *receives information from throughout the interior:* See studies of interoception and the insula: Hugo D. Critchley, "The Human Cortex Responds to Interoceptive Challenge," *Proceedings of the National Academy of Sciences* 101, no. 17 (2004): 6333–34; Hugo Critchley et al., "Neural Systems Supporting Interoceptive Awareness," *Nature Neuroscience* 7 (2004): 189–95; and A. D. (Bud) Craig, "Interoception: The Sense of the Physiological Condition of the Body," *Currrent Opinion in Neurobiology* 13, no. 4 (2003): 500–5. To give a feeling for this science of interoception and our subjective experience, here is a fascinating quote from the abstract of this article: "The primary interoceptive representation in the dorsal posterior insula engenders distinct highly resolved feelings from the body that include pain, temperature, itch, sensual touch, muscular and visceral sensations, vasomotor activity, hunger, thirst, and 'air hunger.' In humans, a meta-representation of the primary interoceptive activity is engendered in the right anterior insula, which seems to provide the basis for the subjective image of the material self as a feeling (sentient) entity, that is, emotional awareness."

38 *As a neuroscientist once said:* An exhibit at the Los Angeles County Museum of Science in the mid-1990s attributed this quote to John Eccles. However, we cannot locate the definitive source.

42 *Researchers have discovered that early experiences:* The field of epigenetics reveals how early experiences directly shape the way genes are regulated. *Epigenetics* refers to the manner in which experience produces neural firing, which in turn selectively turns certain genes "on" and others "off" by way of changes in the chemical controls in the nuclei of the neuronal cells. The net result is to alter the way neurons grow in specific regions of the brain—creating long-lasting structural changes following an experience. Michael Meaney's recent work shows that in people exposed to severe stress early in life, specific genes are activated that continue to shape neural growth throughout the child's life into adulthood. This study examined brain tissue of suicide victims, comparing those who had been abused in childhood with those who had not. Abuse was shown to affect the production of a receptor known to be involved in the stress response. The gene for this glucocorticoid (cortisol) receptor was found to be decreased—a change that is thought to

diminish the control of the response to stress. This lowered number of receptors would make the internal life of the person abused in childhood more stressful. This finding supports the view that experiential factors directly alter the expression of genes—the important process of epigenetics. See Patrick O. McGowan et al., "Epigenetic Regulation of the Glucocorticoid Receptor in Human Brain Associates with Childhood Abuse," *Nature Neuroscience* 12 (2009): 342–48. For positive effects, see Michael J. Meaney, "Maternal Care, Gene Expression, and the Transmission of Individual Differences in Stress Reactivity Across Generations," *Annual Review of Neuroscience* 24 (2001): 1161–92.

44 **Oxytocin is released:** See Thomas R. Insel and Larry J. Young, "The Neurobiology of Attachment," *Nature Reviews: Neuroscience* 2 (2001): 129–36; and Sue Carter, "Neuroendocrine Perspectives on Social Attachment and Love," *Psychoneuroimmunology* 23, no. 8 (November 1998): 779–818. For a discussion of how early experiences shape the oxytocin system, see Alison B. Wismer Fries et al., Early Experience in Humans Is Associated with Changes in Neuropeptides Critical for Regulating Social Behavior," *Proceedings of the National Academy of Sciences* 102, no. 47 (2005): 17237–240.

CHAPTER 3: LEAVING THE ETHER DOME

50 **Medicine too has progressed:** Examples include Columbia University's Program in Narrative Medicine, the University of Rochester's Mindfulness Practice Curriculum, the Harvard program on teaching empathy to medical students, and a program at UCLA teaching first-year students about mindfulness and about the doctor-patient relationship.

52 **Here is the definition:** While an essential aspect of the mind is its regulatory function, naturally our mental experience is filled with layers of inner processes such as our subjective sense of living and our experience of conciousness. In many ways, patterns of energy and information flow enable us to know, perceive, and feel the unique quality of what it means to be alive.

52 **Information *is anything that symbolizes*:** This is a standard "cognitive science" view of information processing. See, for example, Gazzaniga, ed., *The Cognitive Neurosciences,* and Daniel J. Siegel, "Perception and Cognition," in Benjamin Sadock and Virginia Sadock, eds., *Kaplan & Sadock's Comprehensive Textbook of*

Psychiatry, vol. 1, 6th ed. (New York: Lippincott Williams & Wilkins, 1995). See also Evan Thompson, *Mind in Life: Biology, Phenomenology and the Sciences of Mind* (Cambridge, Mass.: Harvard University Press, 2007).

55 *"interpersonal neurobiology":* This field examines the parallel findings from independent disciplines to uncover their common principles. It turns out that this process has a name, which E. O. Wilson describes in the book *Consilience—the Unity of Knowledge* (New York: Vintage, 1998). In his view, consilience enables us to push the boundaries of our knowledge forward by moving beyond the usual constraints of academic fields' often isolated attempts to describe reality. Interpersonal neurobiology is a consilient view that attempts to find these parallel discoveries across numerous ways of knowing—from science, the arts, and contemplative and spiritual practice. In this way, interpersonal neurobiology is not a branch of neuroscience—it is not the same, for example, as social neuroscience. Instead, this field is an open forum for all ways of knowing to collaborate in deepening and expanding our way of understanding reality, the human mind, and well-being.

59 *In the mid-1990s:* See Iacoboni, *Mirroring People;* Laurie Carr et al., "Neural Mechanisms of Empathy in Humans: A Relay from Neural Systems for Imitation to Limbic Areas," *Proceedings of the National Academy of Sciences* 100, no. 9 (2004): 5497–502. The role of mirror neurons and the insula are discussed with regard to their relationship to mindful awareness in Siegel, *The Mindful Brain.* Marco Iacoboni and I presented an all-day discussion of the clinical implications of mirror neurons in 2005. See also Jennifer H. Pfeifer et al., "Mirroring Others' Emotions Relates to Empathy and Interpersonal Competence in Children," *NeuroImage* 39, no. 4 (February 2008): 2076–85.

For an elaboration of the important and complex issue of the role of "dysfunctional" mirror neurons in autism and related disorders, it is important to note that *dysfunctional* literally implies that they are not functioning "normally," and that this could be due to any number of reasons. If children do not find face-to-face interactions safe or interesting, they may "shut off" mirror neuron functions. In other words, this mirror neuron system can be intact without being engaged. Thus an alternative perspective is expressed in the view that the reward system for utilizing social perception is diminished in those with autism and related disorders. In the work of Susan

Bookheimer, for example, decreased activity in the reward-related ventral tegmental region along with diminished orbitofrontal firing was interpreted as the source of decreased reward for attending to social stimuli. This could be a finding supporting the possibility that mirror neurons are intact but the motivational drive to engage socially with others is markedly diminished. Bookheimer reported these findings at the FPR-UCLA Center for Culture, Brain, and Development in a talk entitled "Brain Imaging of Reward Processing and Its Relation to Social Cognition" on February 11, 2009. Supporting this notion that the mirror neuron system is influenced by motivational states is Yawei Cheng, Andrew N. Meltzoff, and Jean Decety, "Motivation Modulates the Activity of the Human Mirror-Neuron System," *Cerebral Cortex* 17, no. 8 (2007): 1979–86.

61 *I once organized an interdisciplinary:* Thanks to the Attias Family Foundation, neuroscientists, anthropologists, developmental psychologists, and those studying psychopathology were all able to gather together for three days of discussion.

62 *the "resonance circuits":* This is fully described in Appendix IIIC of Siegel, *The Mindful Brain.*

62 *The insula brings the resonating state:* See Iacoboni's discussion of the role of the insula in empathy in *Mirroring People,* and Carr et al., "Neural Mechanisms of Empathy."

62 *How, then, do we discern:* Iacoboni, *Mirroring People,* describes a set of neurons he calls the "super mirror neurons," which determine when mirror neurons fire. These are mostly located in the middle prefrontal areas (plus the connected supplementary premotor area) and, along with the increase of input from our own bodies into the cortex via a region called the precuneus, let us know when the mind we are sensing is our own—or that of someone else. These proposed super mirror neurons can also prevent us from imitating or resonating with others when that action or feeling is not appropriate—they may create the foundation for how we distinguish self from other, in Iacoboni's view. Perhaps these super mirror neurons are overactive in the Ether Dome mind state, keeping us from resonating with others, making us feel disconnected and numb. This issue needs further exploration.

CHAPTER 4: THE COMPLEXITY CHOIR

64 *Positive psychology has offered:* See Martin Seligman, *Authentic Happiness* (New York: Free Press, 2002); Martin E. P. Seligman et al.,

"Positive Psychology Progress: Empirical Validation of Interventions," *American Psychologist* 60, no. 5 (2005): 410–21; Sonja Lyubomirsky, *The How of Happiness* (New York: Penguin, 2007).

65 **And it's fascinating to me:** Please see the discussion in Daniel J. Levitan's *This Is Your Brain on Music* (New York: Penguin, 2006).

67 **Take for example the various scientific fields:** See Siegel, *The Developing Mind,* and Daniel J. Siegel, "Emotion as Integration," in Fosha, Siegel, and Solomon, eds. *The Healing Power of Emotion.*

68 **Diving again into the scientific literature:** For a more detailed discussion of the systems views of complexity and chaos theories, see, for example, J. A. Scott Kelso, *Dynamic Patterns: The Self-Organization of Brain and Behavior* (Cambridge, Mass.: MIT Press, 1995); David Bohm, *Wholeness and the Implicate Order* (London: Routledge, 1980); John Holte, ed., *Chaos: The New Science* (Lanham, Md.: University Press of America / The Nobel Conferences, 1990); Stuart Kauffman, *Reinventing the Sacred* (New York: Basic Books, 2008) and *At Home in the Universe: Self-Organization and Complexity* (Oxford: Oxford University Press, 1995). Other useful texts are Ivan Soltesz, *Diversity in the Neuronal Machine: Order and Variability in Interneuronal Microcircuits* (Oxford: Oxford University Press, 2006); and Paul Thagard, *Coherence in Thought and Action* (Cambridge, Mass.: MIT Press, 2000). The role of integration in the creation of consciousness is explored in detail in Gerald M. Edelman and Giulio Tononi, *A Universe of Consciousness: How Matter Becomes Imagination* (New York: Basic Books, 2001).

69 **A system that moves towards complexity:** See Kauffman, *Reinventing the Sacred* and *At Home in the Universe;* Edelman and Tononi, *A Universe of Consciousness;* Bohm, *Wholeness and the Implicate Order.* For a discussion of integration and synchrony in the nervous system's development and function see Marc D. Lewis, "Self-Organizing Individual Differences in Brain Development," *Developmental Review* 25, nos. 3–4 (2005): 252–77. Also see Ulman Lindenberger, Sho Chen Li, Walter R. Gruber, and Viktor Muller, "Brains Swinging in Concert: Cortical Phase Synchronization While Playing Guitar," (BioMedCentral) Neuroscience 10, article 22 (2009); and see Evan Thompson and Francisco J. Varela, "Radical Embodiment; Neural Dynamics and Consciousness," *Trends in Cognitive Neuroscience* 5, no. 10 (2001): 418–25. For a broad scientific and philosophical discussion, see Pier Luigi: Luisi, *Mind and Life* (New York: Columbia University Press, 2004)—especially presentations by Luisi, Michel Bitbol, and

Arthur Zajonc. I would like to acknowledge these three scientists and other co-faculty, including Eshel Ben-Jacob, Fritjof Capra, Nicholas Humphrey, and Stuart Kauffman, for stimulating and affirming discussions regarding integration and self-organization in complex systems. These discussions took place at the Fetzer Institute/Roma 3 International Conference on Science and Spirituality in Cortona, Italy, in June 2009. One realization to emerge from those illuminating conversations was that the essential concept of integration as the linkage of differentiated parts was valid in adaptive self-organization—but the actual term *integration* was generally not used in mathematics or physics because in those fields it means "summation" (e.g., the integration of 3 and 5 is 8). In plain everyday language, however, we can appropriately use *integration*. Integration creates more than the sum of its parts as differentiated elements are linked to one another and complexity increases with adaptive self-organization.

69 **Could it be that mental health was:** In looking for neural correlates of mental health, we don't yet have the published studies that might look at, for example, temporal and spatial neural synchrony as measured using various brain imaging techniques that would correlate with integration. If future studies were to be able to harness technology to assess these neural signatures of a healthy mind, we might be able to see if they were robust factors associated with both the absence of mental disorder and the presence of mental well-being. We would be looking for forms of neural activity that would help us peer into the functioning brain and represent the outcome of the proposal that neural integration, a coherent mind, and empathic relationships are mutually supportive, interactive, and fundamental elements of our triangle of well-being.

CHAPTER 5: A ROLLER-COASTER MIND

80 **This focused attention permits:** The study of consciousness itself is a broad and fascinating pursuit. For an overview of the many research approaches, see "Toward a Science of Consciousness," *Journal of Consciousness Studies,* from the April 2008 conference of the Center for Consciousness Studies at the University of Arizona, Tucson. See also Edelman and Tononi, *A Universe of Consciousness;* Antonio Damasio, *The Feeling of What Happens: The Body and Emotion in the*

Making of Consciousness (New York: Harcourt, 1999); V. S. Ramachandran, *A Brief Tour of Human Consciousness: From Impostor Poodles to Purple Numbers* (New York: Pearson Education, 2004).

80 **The term mood refers:** In a fascinating discussion with Richard Davidson at the Tanner Lecture Series at the University of Utah in February 2009, we explored the nature of "emotion" and the ways in which our evolutionarily older subcortical regions work in concert with our cortex to create what Davidson defines as a "valenced mental state." This useful working definition of emotion enables us to see that appraisal—the determination of whether something is good, neutral, or bad—shapes our overall state of mind. Davidson feels it is important to realize that emotion is an overarching process, not just something created in one region or influencing only an isolated part of the nervous system. For emotion regulation, we can look towards prefrontal function for some insights as this region sends inhibitory fibers downward, through an area called the uncinate fasciculus, to the lower regions. Using new "diffusion tensor imaging" in his lab in Madison, Wisconsin, Davidson and colleagues have demonstrated that this region is a part of how we come to use prefrontal function to regulate subcortical firing. Mindfulness research may reveal how training the mind in focused attention and open monitoring may promote stabilization of our emotional states and the strengthening of our capacity for affect regulation by way of harnessing these prefrontal functions. R. Davidson, personal communication, May 2009.

80 **In a psychiatric textbook:** See Benjamin Sadock and Virginia Sadock, *Kaplan & Sadock's Synopsis of Psychiatry*.

81 **Adults and adolescents with mania:** See Kay Jamison, *An Unquiet Mind* (New York: Random House, 1995).

81 **One current theory is that people with bipolar:** See Hilary Blumberg et al., "Significance of Adolescent Neurodevelopment for the Neural Circuitry of Bipolar Disorder," *Annals of the New York Academy of Sciences* 1021 (2004): 376–83.

82 **The standard treatment for bipolar:** Studies of how medications can promote neuroplasticity include Paul Carlson et al., "Neural Circuitry and Neuroplasticity in Mood Disorders: Insights for Novel Therapeutic Targets," *NeuroRX* 3, no. 1 (2006): 22–41P; Daniela Tardito et al., "Signaling Pathways Regulating Gene Expression, Neuroplasticity, and Neurotrophic Mechanisms in the Action of

Antidepressants: A Critical Overview," *Pharmacological Reviews* 58 (2006):115–34.

84 **In fact, one of the first studies:** See Lewis R. Baxter et al., "Caudate Glucose Metabolic Rate Changes with Both Drug and Behavior Therapy for Obsessive-Compulsive Disorder," *Archives of General Psychiatry* 49, no. 9 (1992): 272–80.

84 **In addition, in our own pilot study:** See Lidia Zylowska et al., "Mindfulness Meditation Training in Adults and Adolescents with ADHD: A Feasibility Study," *Journal of Attention Disorders* 11, no. 6 (2007): 737–46.

84 **Neuroplasticity is possible:** See the extensive review of the literature in Begley, *Train Your Mind, Change Your Brain,* and Doidge, *The Brain That Changes Itself.*

85 **For example, research has also shown:** See Merav Ahissar and Shaul Hochstein, "Attentional Control of Early Perceptual Learning," *Proceedings of the National Academy of Sciences* 90 (1993): 5718–22. See also Aaron R. Seitz and Takeo Watanabe, "Psychophysics: Is Subliminal Learning Really Passive?," *Nature* 422 (2003): 36, and Geoffrey M. Ghose, "Learning in Mammalian Sensory Cortex," *Current Opinion in Neurobiology* 14 (2004): 513–18.

85 **evidence of brain reshaping:** See Thomas Elbert et al., "Increased Cortical Representation of the Fingers of the Left Hand in String Players," *Science* 270 (1995): 305–7.

85 **Other studies have shown that the hippocampus:** See Eleanor A. Maguire et al., "Navigation-Related Structural Change in the Hippocampi of Taxi Drivers," *Proceedings of the National Academy of Sciences* 97, no. 8 (2000): 4398–4403.

85 **In summary, here is what modern:** See Siegel, *The Mindful Brain,* and Amishi P. Jha, Jason Krompinger, and Michael J. Baime, "Mindfulness Training Modifies Subsystems of Attention," *Cognitive, Affective, and Behavioral Neuroscience* 7, no. 2 (2007): 109–19.

86 **As researchers have defined it:** See Jon Kabat-Zinn, *Coming to Our Senses* (New York: Hyperion, 2004); Shauna Shapiro et al., "Mechanisms of Mindfulness," *Journal of Clinical Psychology* 62, no. 3 (2006): 373–86; Susan L. Smalley and Diana Winston, *Fully Present: The Science, Art and Practice of Mindfulness* (New York: DaCapo Press, forthcoming). For another perspective on mindfulness, see Ellen J. Langer, *The Power of Mindful Learning* (New York:DaCapo Press, 1997), and *Counterclockwise: Mindful Healing and the Power of Possibility* (New York: Ballantine, 2009).

86 **Shortly after I had this realization:** See Sara W. Lazar et al., "Meditation Experience Is Associated with Increased Cortical Thickness," *Neuroreport* 16, no. 17 (2005): 1893–97, and Marc D. Lewis and Rebecca M. Todd, "The Self-Regulating Brain: Cortical-Subcortical Feedback and the Development of Intelligent Action," *Cognitive Development* 22, no. 4 (2007): 406–30.

87 **The adolescent brain itself:** See Nitin Gogtay et al., "Dynamic Mapping of Human Cortical Development During Childhood Through Early Adulthood," *Proceedings of the National Academy of Sciences* 101, no. 21 (2004): 8174–79, and Elizabeth A. Sowell et al., "Adolescent Brain and Cognitive Changes," in Martin Fisher et al., eds., *Handbook of Adolescent Medicine* (Elk Grove Village, Ill.: American Academy of Pediatrics, forthcoming).

87 **the foundation for promoting neuroplasticity:** See Doidge, *The Brain That Changes Itself;* Begley, *Train Your Mind, Change Your Brain;* Louis Cozolino, *The Healthy Aging Brain* (New York: Norton, 2008).

88 **With practice, a mindful state:** Mindfulness can be seen as involving two different aspects of mental functioning. One is a trait—an enduring feature of how our minds work that we call part of our "makeup" or "personality." See Ruth A. Baer et al., "Using Self-Report Assessment Methods to Explore Facets of Mindfulness," *Assessment* 13, no. 1 (2006): 27–45. We can also speak of a "state" of mindfulness, or a state of mindful awareness. See Norman A. S. Farb et al., "Attending to the Present: Mindfulness Meditation Reveals Distinct Neural Modes of Self-Reference," *Journal of Social, Cognitive, and Affective Neuroscience* 2, no. 4 (2007): 248–58; Daniel J. Siegel, "Mindfulness Training and Neural Integration: Differentiation of Distinct Streams of Awareness and the Cultivation of Well-Being," *Journal of Social, Cognitive, and Affective Neuroscience* 2, no. 4 (2007): 259–63. For an application of mindful awareness training in children, see Susan Kaiser Greenland, *The Mindful Child* (New York: Free Press, forthcoming); for an overview, see Shauna Shapiro and Linda Carlson, *The Art and Science of Mindfulness* (Washington, D.C.: APA Press, 2009). See also Jack Kornfield, *The Wise Heart* (New York: Bantam, 2007).

89 **A picture in my own mind:** The "wheel of awareness" was first published in Siegel, *The Mindful Brain.*

89 **Here is a transcript:** I first offered this meditation in a public setting at the Mind and Moment conference in 2006 in San Francisco with Diane Ackerman, Jon Kabat-Zinn, and the late John O'Donohue.

93 *"The faculty of voluntarily"*: See William James, *Principles of Psychology* (Cambridge, Mass.: Harvard University Press, 1981), 401. On mindfulness, see Kirk Warren Brown, Richard M. Ryan, and J. David Creswell, "Mindfulness: Theoretical Foundations and Evidence for its Salutary Effects," *Psychological Inquiry* 18, no. 4 (2007): 211–37; Shapiro and Carlson, *The Art and Science of Mindfulness;* Kabat-Zinn, *Coming to Our Senses.* See also A. Jha, J. Krompinger, and M. J. Blaine, "Mindfulness Training Modifies Subsystems of Attention," *Cognitive, Affective Behavioral Neuroscience* 7 (2007): 109–19.

100 **What had changed:** See the fascinating work of Fred Gage on hippocampal growth with voluntary but not forced physical activity, in Henriette van Praag et al., "Exercise Enhances Learning and Hippocampal Neurogenesis in Aged Mice," *Journal of Neuroscience* 25, no. 38 (2005): 8680–85.

CHAPTER 6: HALF A BRAIN IN HIDING

105 *A vast amount of research suggests:* See Siegel, *The Developing Mind,* and Erik Hesse et al., "Unresolved States Regarding Loss and Abuse Can Have 'Second Generation Effects,' " in Solomon and Siegel, eds., *Healing Trauma.*

106 *Perhaps it was being raised:* See the work on the genetics of personality in Lawrence A. Pervin and Oliver P. John, eds., *Handbook of Personality: Theory and Research,* 2nd ed. (New York: Guilford, 2001), especially Robert Plomin and Avshalom Caspi, "Behavioral Genetics and Personality," 251–76.

106 *To understand Stuart:* See Siegel, *The Developing Mind,* for an in-depth discussion of attachment and brain laterality. For the discussions about differences between the right and left brains, see Eran Zaidel and Marco Iacoboni, eds., *The Parallel Brain: The Cognitive Neuroscience of the Corpus Callosum* (Cambridge, Mass.: MIT Press, 2003); Sally P. Springer and Georg Deutsch, *Left Brain, Right Brain: Perspectives from Cognitive Neuroscience* (New York: Freeman, 1997); and Chris McManus, *Right Hand, Left Hand* (Cambridge, Mass.: Harvard University Press, 2002).

115 **For homework, I gave him a book:** See Betty Edwards, *Drawing on the Right Side of the Brain* (New York: Tarcher/Penguin, 1979).

116 *And indeed, studies done:* See David Creswell et al., "Neural Correlates of Dispositional Mindfulness During Affect Labeling," *Psychosomatic Medicine* 69 (2007): 560–65.

118 *Our focus on his bodily sensations:* See Orin Devinsky, "Right Cerebral Hemisphere Dominance for a Sense of Corporeal and Emotional Self," *Epilepsy and Behavior* 1 (2000): 60–73.

CHAPTER 7: CUT OFF FROM THE NECK DOWN

122 *Mr. Duffy "lived . . .":* James Joyce, *Dubliners* (New York: Signet, 1993).

124 *When I asked if she could sense:* Some suggest that awareness of the heart is a sign of interoceptive capacity. See Antoine Bechara and Nasir Naqvi, "Listening to Your Heart: Interoceptive Awareness as a Gateway to Feeling," *Nature Neuroscience* 7 (2004): 102–3.

125 *Research has shown repeatedly:* See Ran R. Hassin, James S. Uleman, and John A. Bargh, eds., *The New Unconscious* (Oxford: Oxford University Press, 2006), as well as studies of implicit memory that we'll turn to in the next chapter.

125 *Colleagues of mine at UCLA:* See Naomi Eisenberger and Matt Lieberman, "Why Rejection Hurts: A Common Neural Alarm System for Physical and Social Pain," *Trends in Cognitive Sciences* 8, no. 7 (2004): 294–300.

125 *In fact, the more we can sense:* See A. D. (Bud) Craig, "How Do You Feel—Now? The Anterior Insula and Human Awareness," *Nature Reviews: Neuroscience* 10, no. 1 (2009): 59–70; Hugo D. Critchley, "The Human Cortex Responds to an Interoceptive Challenge," *Proceedings of National Academy of Science* 101, no. 17 (2004): 6333–34; Olga Pollatos, Klaus Gramann, and Rainer Schandry, "Neural Systems Connecting Interoceptive Awareness and Feelings," *Human Brain Mapping* 28, no. 1 (2007): 9–18; Hugo D. Critchley, "Neural Mechanisms of Autonomic, Affective, and Cognitive Integration," *Journal of Comparative Neurology* 493 (2005):154–66; Hugo D. Critchley et al., "Neural Systems Supporting Interoceptive Awareness," *Nature Neuroscience* 7 (2004): 189–95; A. D. (Bud) Craig, "How Do You Feel? Interoception: The Sense of the Physiological Condition of the Body," *Nature Reviews: Neuroscience* 3 (2002): 655–66; Tania Singer et al., "Empathy for Pain Involves the Affective but not Sensory Components of Pain," *Science* 303 (2004): 1157–62; A. D. (Bud) Craig, "Human Feelings: Why Are Some More Aware than Others?," *Trends in Cognitive Sciences* 8, no. 6 (2004): 239–41.

126 *The insula and ACC:* The anterior insula and another area of the middle prefrontal region, the anterior cingulate, share a unique form of

cell, the "von Economo neuron," otherwise known as the spindle cell. These are long, highly interconnecting cells that are exclusively located in the resonance circuits. One view is that these cells may create fast communication among the physically distant areas, such as between the anterior insula and cingulate. Spindle cells are most numerous in mature people and then are progressively less numerous in children, gorillas, and then chimps. In monkeys and most other mammals, they do not even exist. These patterns of spindle cell populations parallel the distribution of the ability for self-recognition, leading some researchers to suggest that these unusual cells may play an important role in self-awareness. Two non-primate mammals also have the ability to recognize themselves in a mirror (the standard "self-awareness test"): elephants and dolphins. It turns out that they, too, have spindle cells.

With our highly integrative spindle-cell-rich insula and anterior cingulate, we create an awareness of our emotional state that lets us monitor our internal world. With our related mirror neuron functions, we also create an awareness of others' internal experience. But the insula-cingulate connection permits yet another capacity. The anterior cingulate works closely with other aspects of the middle prefrontal cortex to modify our internal states. If people find a way to shut off flow to the anterior insula and anterior cingulate, they'll eliminate not only awareness but also the ability to shape their emotional state. They'll lack the capacity to monitor and modify their internal world with their cortex.

126 *The brainstem also works:* See the work of Porges, who describes the polyvagal theory. Here different branches of the vagal nerve and the sympathetic branch of the autonomic nervous system participate in our brainstem's response to threat. He has coined a term, *neuroception,* which denotes the way we evaluate a situation for threat and then activate the fight-flight-freeze response if we feel in danger. If we assess safety, we turn on the "social engagement" system and become receptive to others. Porges calls this "love without fear." Inspired by his work, I've described a parallel process (in *The Mindful Brain*) in which internal attunement creates a state of safety and then turns on our self-engagement system—we become open to ourselves, ready to become our own best friend. See Porges, "Reciprocal Influences Between Body and Brain in the Perception and Expression of Affect."

128 *If we focus only on the easily named:* See Paul Ekman and Erika Rosenberg, *What the Face Reveals: Basic and Applied Studies of*

Spontaneous Expression Using the Facial Action Coding System (FACS), 2nd ed. (Oxford: Oxford University Press, 2005).

128 ***Primary emotion is the subtle music:*** See Siegel, *The Developing Mind*, for discussion of the concept of primary emotion.

132 ***It was only later that:*** See Michael Anderson's studies of repression and brain function. Michael Anderson et al., "Neural Systems Underlying the Suppression of Unwanted Memories," *Science* 9, no. 303 (2004): 232–35.

133 ***The nucleus basalis:*** See A. A. Miasnikov, J. C. Chen, and N. M. Weinberger, "Behavioral Memory Induced by Stimulation of the Nucleus Basalis: Effects of Contingency Reversal," *Neurobiology of Learning and Memory* 91, no. 3 (2009): 298–309, and A. A. Miasnikov et al., "Motivationally Neutral Stimulation of the Nucleus Basalis Induces Specific Behavioral Memory," *Neurobiology of Learning and Memory* 90, no. 1 (2008): 125–37.

134 ***While research suggests that:*** See extensive data at Heartmath .com; and see Bechara and Naqvi, "Listening to Your Heart."

135 ***This technique is used:*** See Ogden, Pain, and Minton, *Trauma and the Body*, and Peter Levine, *Waking the Tiger* (Berkeley, Calif.: North Atlantic, 1997).

136 ***Still others involve bilateral stimulation:*** EMDR, or Eye Movement Desensitization and Reprocessing, is an approach to therapy that involves a protocol enabling various sensations, images, and thoughts to be brought together along with bilateral stimulation to facilitate change. Francine Shapiro's *EMDR*, 2nd ed. (New York: Guilford, 2001) and her edited volume *EMDR as an Integrative Psychotherapy Approach: Experts of Diverse Orientations Explore the Paradigm Prism* (Washington, D.C.: APA Press, 2002) are good overviews.

CHAPTER 8: PRISONERS OF THE PAST

145 ***Our paths came together:*** See Solomon and Siegel, eds., *Healing Trauma*.

146 ***I'd learned about our ability:*** See Solomon and Siegel, eds., *Healing Trauma*, and Van der Hart, Nigenhuis, and Steele, *The Haunted Self*.

147 ***In the years since my encounter:*** For an overview, see Bessel van der Kolk, "Posttraumatic Stress Disorder and the Nature of Trauma," in Solomon and Siegel, eds., *Healing Trauma*, 168–95.

147 **overwhelms their ability to cope:** See Ogden, Pain, and Minton, *Trauma and the Body.*

147 **Memory is the way an experience:** See Daniel J. Siegel, "Memory: An Overview with Emphasis on the Developmental, Interpersonal, and Neurobiological Aspects," *Journal of the American Academy of Child and Adolescent Psychiatry* 40 (2000): 997–1011.

147 **The gene activation and protein production:** See Doidge, *The Brain That Changes Itself,* and Begley, *Train Your Mind, Change Your Brain.*

148 **It can also thicken the insulating:** Myelin is the fatty sheath that serves as insulation allowing the ion flow—the equivalent of an electric current—to increase its speed one hundred times. The region where synapses are found in the cortex is without myelin and is a gray color, but when myelin covers the long axonal lengths, they are whitish, and thus this axonal region is called the white matter. For this general topic, please see the reviews by Doug Field, "White Matter Matters," *Scientific American,* March 2008, 54–61, and "Myelination: An Overlooked Mechanism of Synaptic Plasticity?," *Neuroscientist* 11, no. 6 (2005): 528–531. Regarding the research into skills and myelin growth, please see E. M. Miller's "Intelligence and Brain Myelination: A Hypothesis," *Personality and Individual Differences* 17 (1994): 803–32. Also F. Ullen and colleagues' study of piano practice in "Extensive Piano Practicing Has Regionally Specific Effects on White Matter Development," *Neuroscience* 8 (2005): 1148–50.

148 **Neurons that fire together, wire together:** This phrase is generally attributed to Donald Hebb, a Canadian doctor and psychologist whose 1949 book, *The Organization of Behavior: A Neuropsychological Theory,* postulates this notion that neurons firing simultaneously at one time will be more likely to fire together in the future. This associational linkage underlies the term *Hebbian synapse* that refers to increased connectivity among previously firing neurons. Norman Doidge attributes the actual wording to Carla Shatz, but notes that in fact Sigmund Freud in 1888 made similar suggestions (which he called the "law of association by simultaneity"). Virtually all of the research on memory has confirmed that Hebb's and Freud's intuitions and propositions were correct. For example, Eric Kandel, a psychiatrist, explored this notion in the sea slug and found the basics of learning—and then in 2000 received the Nobel Prize for his crucial contributions. See Kandel, *In Search of Memory.*

148 *Here's a key fact about memory retrieval:* See, for example, Daniel
Schacter, *Searching for Memory: The Brain, the Mind, and the Past*
(New York: Basic Books, 1996), and Larry Squire and Daniel
Schacter, *Neuropsychology of Memory*, 3rd ed. (New York: Guilford,
2003); Kandel, *In Search of Memory.*

151 *If you had been a volunteer:* On the "dichotic listening experiment,"
see Lutz Jancke et al., "Focused Attention in a Simple Listening
Task: An fMRI Experiment," *Cognitive Brain Research* 16, no. 2 (2003):
257–66.

151 *Direct attention harnesses:* See the work of Daniel Schacter, in
Squire and Daniel Schacter, eds., *Neuropsychology of Memory.*

152 *The implicit mental models:* On schema, see Darcia Narvaez and
Tonia Bock, "Moral Schemas and Tacit Judgment, or How the
Defining Issues Test Is Supported by Cognitive Science," *Journal of
Moral Education* 31, no. 3 (2002): 297–314. Also see Phillip Johnson
Laird, "Inference and Mental Models," in Stephen Newstead and
Jonathan Evans, eds., *Perspectives on Thinking and Reasoning*
(Mahwah, N.J.: Erlbaum, 1994); William A. Cunningham and Phillip
David Zelazo, "Attitudes and Evaluations: A Social Cognitive
Neuroscience Perspective," *Trends in Cognitive Sciences* 11, no. 3
(2007): 97–104.

153 *Explicit memory begins to emerge:* See Carolyn K. Rovee-Collier,
Harlene Hayne, and Michael Colombo, *The Development of Implicit
and Explicit Memory* (Amsterdam and Philadelphia: John Benjamins,
2001).

155 *Rage can also shut off:* With excessive stress, the hormone cortisol
leads to inhibition of normal hippocampal function and growth. See
Robert M. Sapolsky, "Glucocorticoids and Hippocampal Atrophy in
Neuropsychiatric Disorders," *Archives of General Psychiatry* 57 (2000):
925–35. See Larry R. Squire and Stuart Zola-Morgan, "The Medial
Temporal Lobe Memory System," *Science* 253 (1991): 1380–86, for a
general review of the hippocampus from early studies, and Squire
and Schacter, ed., *Neuropsychology of Memory.* Another set of as yet
unpublished studies reveals that children raised during the first years
of their lives in institutional settings, such as orphanages, have a number of findings thought to be due to excessive stress from the rigid,
unpredictable, at times neglectful environment of that setting. These
findings include a larger amygdala—and at times a smaller hippocampus. The degree of enlargement of the amygdala corresponded to

the amount of emotional confusion these children would experience when shown photographs of negatively valenced faces. Of note, too, was that the larger amygdala was also associated with decreased focus on the eye region of the face. In these ways, a cascade of developmental stress could be seen to unfold in this manner: Environmental stress → increased amygdala growth → increased emotional reactivity to negative emotional facial expression *and* decreased perception of facial features. The result of this unfortunate situation was proposed to be difficulties with a) emotional regulation, b) self-organization in social settings, and c) decreased perceptual experience of seeing faces. Of note also was that when noninstitutionalized children saw faces, they used their cortical regions (including the superior temporal cortex and the fusiform gyrus, the latter being involved in expertise), whereas those raised in institutions did not activate these higher regions but instead had the stimulation of the amygdala and other subcortical areas. All of these findings suggest that the way children will remember experiences— both implicitly and explicitly—will be shaped by the early years of life. These results were presented by Nim Tottenhan, Ph.D., at the Foundation for Psychocultural Research–UCLA Center for Culture, Brain, and Development's colloquium in a talk entitled "Neuro-Behavioral Development Following Early Life Stress" on February 25, 2009.

156 **When I first read this research:** The first presentation I made on the role of the hippocampus in trauma was at the meeting of the American College of Psychiatrists in San Francisco in 1992, which had the conference theme "Memories: True, False, and Absent." The ideas of that workshop were published as "Memory, Trauma, and Psychotherapy: A Cognitive Science View," *Journal of Psychotherapy Practice and Research* 4, no. 2 (1995): 93–122. This view was elaborated in Siegel, *The Developing Mind;* Marian Sigman and Daniel J. Siegel, "The Interface Between the Psychobiological and Cognitive Models of Attachment," *Behavioral and Brain Sciences* 15, no. 3 (1992): 523; Theodore Gaensbauer et al., "Traumatic Loss in a One-Year-Old Girl," *Journal of Child and Adolescent Psychiatry* 34, no. 4 (1995); Daniel J. Siegel, "Dissociation, Psychotherapy, and the Cognitive Sciences, in James L. Spira, ed., *Treating Dissociative Identity Disorder* (San Francisco: Jossey-Bass, 1995), 39–79; "Cognition, Memory, and Dissociation," in Dorothy O. Lewis and Frank W. Putnam, eds., *Child and Adolescent Psychiatric Clinics of North America on Dissociative*

Disorders (Philadelphia: W. B. Saunders, 1996); and Daniel J. Siegel, "Toward an Interpersonal Neurobiology of the Developing Mind: Attachment, 'Mindsight,' and Neural Integration," *Infant Mental Health Journal* 22 (2001): 67–94.

157 **High levels of adrenaline act to:** See Bennet M. Elzinga and James D. Brenner, "Are the Neural Substrates of Memory the Final Common Pathway in Posttraumatic Stress Disorder (PTSD)?," *Journal of Affective Disorders* 70, no. 1 (2002): 1–17.

158 **Sleep phenomena such as nightmares:** See Thomas A. Mellman et al., "REM Sleep and the Early Development of Posttraumatic Stress Disorder," *American Journal of Psychiatry* 159 (2002): 1696–1701; Giora Pillar, Atul Malhotra, and Peretz Lavie, "Post-Traumatic Stress Disorder and Sleep—What a Nightmare!," *Sleep Medicine Reviews* 4, no. 2 (2000): 183–200.

161 **When families do not offer:** See Siegel and Hartzell, *Parenting from the Inside Out,* for a practical approach to making sense of early life experience.

162 **My old memory mentor:** Robert Bjork has made major contributions to our understanding of learning. See his "Memory and Metamemory: Considerations in the Training of Human Beings," in Janet Metcalfe and Arthur P. Shimamura, eds., *Metacognition: Knowing About Knowing* (Cambridge, Mass.: MIT Press, 1994), 185–205.

CHAPTER 9: MAKING SENSE OF OUR LIVES

166 **It wasn't until years later:** See Sroufe, Egeland, and Carlson, *The Development of the Person.* See also Eisenberger and Lieberman, "Why Rejection Hurts," Critchley et al., "Neural Systems," and Bechara and Naqvi, "Listening to Your Heart."

167 **For me, the explanation lies:** See Jude Cassidy and Phil Shaver, eds., *Handbook of Attachment* (New York: Guilford, 1999). For the ensuing discussion of the attachment paradigm in infants (the Infant Strange Situation) and in adults (the Adult Attachment Interview) see the relevant sections of this work and Sroufe, Egeland, and Carlson, *The Development of the Person.*

168 **About two-thirds of children:** These statistics are for the U.S. population. Statistics can vary by the culture being studied and when the studies were carried out. New findings can change, and studying high-risk populations—such as those living in poverty, with drug addiction, or with other challenges to mental health—may reveal

quite different degrees of nonsecure attachment. Also, anthropologists have suggested that we avoid "pathologizing" the subjects; "insecure" attachment may be derogatory to the child. In other words, the "problem" is not with the child being an "insecure" person but rather that the relationship was suboptimal and hence not secure.

168 *Another 10 to 15 percent:* Another term for this grouping is *resistant* attachment, in that the child resists being comforted in the Infant Strange Situation.

170 *Studies have indeed established:* See Robert Plomin et al., *Behavioral Genetics,* 4th ed. (New York: Worth, 2000).

170 *One of the leading researchers:* This was the spontaneous offering from Robert Plomin at the American Psychiatric Association Annual Meeting, New York, May 2004. A discussion of this issue of genetics and attachment can be found in these two articles: Kathryn A. Becker-Blease et al., "A Genetic Analysis of Individual Differences in Dissociative Behaviors in Childhood and Adolescence," *Journal of Child Psychology and Psychiatry* 45, no. 3 (2004): 522–32, and Caroline L. Bokhorst et al., "The Importance of Shared Environment in Mother–Infant Attachment Security: A Behavioral Genetic Study," *Child Development* 74, no. 6 (November/December 2003): 1769–82. Below are two thoughtful discussions of how the child's environment plays a major role in determining attachment outcome. Of note in the first paper is the finding that the genetic contribution to absorption—or "normal dissociation"—may be high and that the exposure of that child to psychological unavailability or terror can then induce pathological dissociation. A related article of note is the finding that children with a genetic variant of their dopamine circuitry may have more intense responses to overwhelming events. Here are those references: Marian J. Bakermans-Kranenburg and Marinus H. van Ijzendoorn, "Research Review: Genetic Vulnerability or Differential Susceptibility in Child Development: The Case of Attachment," *Journal of Child Psychology and Psychiatry* 48, no. 12 (2007): 1160–73; Krisztina Lakatos et al., "Further Evidence for the Role of the Dopamine D4 Receptor (DRD4) Gene in Attachment Disorganization: Interaction of the Exon III 48-bp Repeat and the 521 C/T Promoter Polymorphisms," *Molecular Psychiatry* 7, no. 1 (2002): 27–31.

171 *Furthermore, research with foster:* See Mary Dozier et al., "Attachment for Infants in Foster Care: The Role of Caregiver State of Mind," *Child Development* 72 (2001): 1467–77.

171 *But anyone who doubts the influence:* For a contrasting perspective, and a reminder that not only are peers and genes important, but

parents do not determine all of development, see Judith Rich Harris, *The Nurture Assumption* (New York: Free Press, 1998). See also the foreword in that book, by Steven Pinker, as well as his position regarding the overemphasis in modern thinking on the influence of parents on children, in *How the Mind Works* (New York: Norton, 1997). While genetics is important in temperament, temperament and genetics are not predominant influences on attachment categories but experience with the caregivers is. See Brian Vaughn and Kelly Bost, "Attachment and Temperament," in Cassidy and Shaver, eds., *Handbook of Attachment*, 198–225; Marian Bakermans-Kranenburg et al., "The Importance of Shared Environment in Mother-Infant Security," *Child Development* 74, no. 6 (2003): 1769–82.

173 **The research instrument:** See Hesse et al., "Unresolved States," in Howard Steele and Miriam Steele, eds., *Clinical Applications of the Adult Attachment Interview* (New York: Guilford, 2008), and Mary Main, "The Adult Attachment Interview: Fear, Attention, Safety, and Discourse Processes," *Journal of the American Psychoanalytic Association* 48 (2000): 1055–96.

175 **Patients with coherent narratives:** The formal research measure of a parallel process to mindsight is called *mentalization*, otherwise known as *reflective function*. This process has predecessors in the academic literature with names such as *theory of mind*, *psychological-mindedness, mind-mindedness*, and *mentalese*. See Allen, Fonagy, and Bateman, *Mentalizing in Clinical Practice*.

175 **As a touchstone for our discussion:** Abstracted from Siegel, *The Developing Mind*, 70.

178 **Attachment researchers have monitored:** These studies can be found in the *Handbook of Attachment*. See also R. Chris Fraley, Keith E. Davis, and Philip R. Shaver, "Dismissing-Avoidance and the Defensive Organization of Emotion, Cognition, and Behavior," in Jeffrey A. Simpson and William Rholes, eds., *Attachment Theory and Close Relationships* (New York: Guilford, 1997), 249–79; Mary Dozier et al., "The Challenge of Treatment for Clients with Dismissing States of Mind," *Attachment and Human Development* 3, no. 1 (2001): 62–76.

185 **When the same person:** This is the formulation from Mary Main and Erik Hesse of the biological paradox that leads to "fright without solution." See Hesse et al., "Unresolved States."

187 **Studies have suggested:** See James Pennebaker, "Telling Stories: The Health Benefits of Narrative," *Literature and Medicine* 19, no. 1 (2000): 3–18, and *Opening Up: The Healing Power of Expressing Emotions* (New York: Guilford, 1997).

188 *But they can also emerge:* See Siegel and Hartzell, *Parenting from the Inside Out* for applications of these ideas of developing an earned security.

195 *Shame states are common:* See Schore, *Affect Dysregulation and Disorders of the Self.*

196 *It is here that we can begin:* See Porges, "Reciprocal Influences."

197 *At the extreme end of the spectrum:* For a discussion of dissociation in all of its normal and adaptive components as highlighted in this chapter, see Paul Dell and John O'Neil, eds., *Dissociation and the Dissociative Disorders: DSM-V and Beyond* (London: Routledge, 2009), including the chapter by Lissa Dutra et al., "The Relational Context of Dissociative Phenomena," 83–92. See also A. A. T. Simone Reinders et al., "Psychobiological Characteristics of Dissociative Identity Disorder: A Symptom Provocation Study," *Biological Psychiatry* 60, no. 7 (2006): 730–40.

198 *Early adolescence is filled:* See Susan Harter, *The Construction of the Self: A Developmental Perspective* (New York: Guilford, 1999).

198 *In brain terms, a state is composed:* See the chapter on "states of mind" in Siegel, *The Developing Mind.*

200 *Many self-states are organized:* See Panksepp, *Affective Neuroscience* and "Brain Emotional Systems." Panksepp has suggested that we have many subcortically organized and somewhat independently operating motivational drives, such as for play, mastery, resource allocation, reproduction, and caregiving. For a contrasting perspective that emphasizes the importance of the cortex in the experience of emotion, see Richard J. Davidson, "Seven Sins in the Study of Emotion: Correctives from Affective Neuroscience," *Brain and Cognition* 52, no. 1 (2003): 129–32.

201 *To understand how states of mind:* For a review of our six-layered cortex, see Jeffrey Hawkins and Sandra Blakeslee, *On Intelligence* (New York: Times Books, 2004).

207 *I showed him a way to hold himself:* I thank Pat Ogden for demonstrating this technique to me.

208 *Some researchers call this core:* See Siegel, *The Mindful Brain,* for a full discussion of this notion of an "ipseitious self " and the contemplative science view of this core of our internal world. See also Antoine Lutz, John D. Dunne, and Richard J. Davidson, "Meditation and the

Neuroscience of Consciousness: An Introduction," in Philip D. Zelazo, Morris Moscovitch, and Evan Thompson, eds., *The Cambridge Handbook of Consciousness* (Cambridge, U.K.: Cambridge University Press, 2007), 499–554.

CHAPTER 11: THE NEUROBIOLOGY OF "WE"

218 *Even his performance as a jazz pianist:* Interestingly, jazz improvisation requires the middle prefrontal areas to be actively engaged, in contrast to classical performance. See Charles J. Limb and Allen R. Braun, "Neural Substrates of Spontaneous Musical Performance: An fMRI Study of Jazz Improvisation," *PLoS ONE* 3 no. 2 (2008): e1679; doi: 10.1371/journal.pone.0001679.

222 *His mother had been emotionally blunted:* See Geraldine Dawson et al., "Preschool Outcomes of Children of Depressed Mothers: Role of Maternal Behavior, Contextual Risk, and Children's Brain Activity," *Child Development* 74, no. 4 (2003): 1158–75.

222 *it's like living in a chronic "still-face" experiment:* See Tronick, *The Neurobehavioral and Social Emotional Development of Infants and Children.*

223 *"growth edges":* I thank David Daniels, M.D., for introducing me to this term.

224 *Our mirror neuron system "learns":* This essential notion comes from the work of Iacoboni and is built from a range of studies originating in Italy with the work of Giacomo Rizolatti and Vittorio Gallese. See Vittorio Gallese and Alvin Goldman, "Mirror Neurons and the Simulation Theory of Mindreading," *Trends in Cognitive Sciences* 2 (1998): 493–501; Giacomo Rizolatti and Michael A. Arbib, "Language Within Our Grasp," *Trends in Neuroscience* 21 (1998): 188–194; Vittorio Gallese, "Intentional Attunement: A Neurophysiological Perspective on Social Cognition and Its Disruption in Autism," *Brain Research* 1079 (2006): 15–24.

225 *Their limbic system's amygdala fires off:* See P. Vrtička et al., "Individual Attachment Style Modulates Human Amygdala and Striatum Activation During Social Appraisal," *PLoS ONE* 3, no. 8 (2008): e2868; doi: 10.1371/journal.pone.0002868. The striatum is essential in creating motivational drives and was reduced in its activation in those with avoidant histories seeing a smiling face—while the amygdala was increased in activation in response to hostile faces for those with ambivalent histories.

CHAPTER 12: TIME AND TIDES

232 *The adolescent brain changes:* See Sowell, Siegel, and Siegel, "Adolescent Brain and Cognitive Changes"; Sarah-Jayne Blakemore, "The Social Brain in Adolescence," *Nature Reviews: Neuroscience* 9 (2008): 267–77; Gogtay et al., "Dynamic Mapping of Human Cortical Development."

233 *We now know that some of our fellow mammals:* See G. A. Bradshaw et al., "Elephant Breakdown," *Nature* 433 (2005): 807.

238 *A wide variety of cognitive experiments:* See Jennifer Freyd's work on dynamic representations: "Dynamic Mental Representations," *Psychological Review* 94 (1987): 427–38; "Five Hunches About Perceptual Processes and Dynamic Representations," in David E. Meyer and Sylvan Kornblum, eds., *Attention and Performance XIV: Synergies in Experimental Psychology, Artificial Intelligence, and Cognitive Neuroscience* (Cambridge, Mass.: MIT Press, 1993), 99–119.

241 *OCD can come on suddenly:* See Susan E. Swedo, Henrietta L. Leonard, and Judith L. Rapoport, "The Pediatric Autoimmune Neuropsychiatric Disorders Associated with Streptococcal Infection (PANDAS) Subgroup: Separating Fact from Fiction," *Pediatrics* 113, no. 4 (2004): 907–11.

241 *Some doctors who diagnose OCD:* Selective serotonin reuptake inhibitors (SSRIs) are a common class of medications used to treat OCD. In this article, you'll see the shared view that medications should not be a first-line treatment, especially in children or adolescents. See I. Heyman, D. Mataix-Cols, and N. A. Fineberg, "Obsessive-Compulsive Disorder," *British Medical Journal* 333 (2006): 424–29. For further research on OCD, see S. P. Whiteside, J. D. Port, and J. S. Abramowitz, "A Meta-Analysis of Functional Neuroimaging in Obsessive-Compulsive Disorder," *Psychiatry Research* 132 (2004): 69–79; K. Richard Ridderinkhof et al., "The Role of the Medial Frontal Cortex in Cognitive Control," *Science* 306, no. 5695 (2004): 443–47; James Woolley et al., "Brain Activation in Paediatric Obsessive-Compulsive Disorder During Tasks of Inhibitory Control," *British Journal of Psychiatry* 192 (2008): 25–31.

242 *Research with adults had shown:* One of the first studies to demonstrate long-term changes in symptom relief and brain function was performed at UCLA and used cognitive-behavioral strategies, discussions of the brain, and mindfulness as one component of the treatment with adults. See Baxter et al., "Caudate Glucose Metabolic Rate Changes."

EPILOGUE

255 *In 1950, Albert Einstein:* This letter was quoted years later in *The New York Times* (March 29, 1972) and the *New York Post* (November 28, 1972). I thank Jon Kabat-Zinn for introducing me to these words; see his book *Full Catastrophe Living* (New York: Delta, 1990), 166.

257 *For as long as we have had records:* See Jeffrey Moses, *Oneness: Great Principles Shared by All Religions,* revised and expanded edition (New York: Random House, 2002).

257 *Today we can actually track:* Numerous approaches help us elucidate the nature of the self and neural functions. See Jason P. Mitchell, Mahzarin R. Banaji, and C. Neil Macrae, "The Link Between Social Cognition and Self-Referential Thought in the Medial Prefrontal Cortex," *Journal of Cognitive Neuroscience* 17, no. 8 (2005): 1306–15; Decety and Moriguchi, "The Empathic Brain and Its Dysfunction."

257 *But if we cannot identify:* See Mitchell, Banaji, and Macrae, "The Link Between Social Cognition and Self-Referential Thought"; Lucina Q. Uddin, Marco Iacoboni, Claudia Lange, and Julian Paul Keenan, "The Self and Social Cognition: The Role of Cortical Midline Structures and Mirror Neurons," *Trends in Cognitive Sciences* 11 (2007): 153–157; Matthew D. Lieberman, "Social Cognitive Neuroscience: A Review of Core Processes," *Annual Review of Psychology* 58 (2007): 259–89; Vittorio Gallese, Christian Keysers, and Giacomo Rizzolatti, "A Unifying View of the Basis of Social Cognition," *Trends in Cognitive Sciences* 8, no. 9 (2004): 396–403.

258 *Imaging studies have demonstrated:* Ahmad R. Hariri et al., "The Amygdala Response to Emotional Stimuli: A Comparison of Faces and Scenes," *NeuroImage* 17, no. 1 (2002): 317–23, and Yi Jiang and Sheng He, "Cortical Responses to Invisible Faces: Dissociating Subsystems for Facial-Information Processing," *Current Biology* 16, no. 2 (2006): 2023–29.

258 *Such "mortality salience" studies:* See Holly McGregor et al., "Terror Management and Aggression: Evidence That Mortality Salience Motivates Aggression Against Worldview-Threatening Others," *Journal of Personality and Social Psychology* 74, no. 3 (1998): 590–605; Susan T. Fiske, "Social Cognition and the Normality of Prejudgment," in John Dovidio, Peter Glick, and Laurie Rudman, eds., *On the Nature of Prejudice* (Malden, Mass.: Wiley Blackwell, 2005); Mario Mikulincer and Victor Florian, "Exploring Individual Differences in Reactions to Mortality Salience: Does Attachment Style Regulate Terror Management Mechanisms?," *Journal of*

Personality and Social Psychology 79, no. 2 (2000): 260–73; Joshua Hart, Phil-lip R. Shaver, and Jamie L. Goldenberg, "Attachment, Self-Esteem, Worldviews, and Terror Management: Evidence for a Tripartite Security System," *Journal of Personality and Social Psychology* 88, no. 6 (2005): 999–1013. See also Samuel Bowles, "Group Competition, Reproductive Leveling and the Evolution of Human Altruism," *Science* 314, no. 5805 (2006): 1569–72. Also see Charles R. Efferson, Rafael Lalive, and Ernst Fehr, "The Coevolution of Cultural Groups and In-Group Favoritism," *Science* 32, no. 5897 (2008): 1844–49; Susan T. Fiske, "What We Know About Bias and Intergroup Conflict, the Problem of the Century," *Current Directions in Psychological Science* 11, no. 4 (2002): 123–28.

259 ***The study of positive psychology:*** Studies reveal that even winning the lottery does not make you happier. Contrary to popular belief, what we think will make us happy and what actually does don't correspond. See Seligman, *Authentic Happiness;* Daniel Gilbert, *Stumbling on Happiness* (New York: Random House, 2006); and Lyubomirsky, *The How of Happiness;* Elizabeth W. Dunn, Lara Baknin, and Michael I. Norton, "Spending Money on Others Promotes Happiness," *Science* 319, no. 5870 (2008): 1687–88. Also see Dacher Keltner, *Born to Be Good* (New York: W. W. Norton, 2009).

261 ***Physically and genetically, our brains:*** These references explore how social factors play a significant role in the evolution of our brain—in size and in complexity. With our genetically shaped potentials, our cultural experiences directly influence how our individual brains develop. See David Lewis-Williams, *The Mind in the Cave* (London: Thames & Hudson, 2002); Steven Mithen, *The Prehistory of the Mind* (London: Thames & Hudson, 1996); Donald Merlin, *A Mind So Rare* (New York: Norton, 2001). For a discussion of cognitive evolution, also see Michael Tomasello, *The Cultural Origins of Human Cognition* (Cambridge, Mass.: Harvard University Press, 1999). Also see Michael Balter, "Brain Evolution Studies Go Micro," *Science* 315 (2007): 1208–11; R. I. M. Dunbar and Suzanne Shultz, "Evolution in the Social Brain," *Science* 317, no. 5843 (2007): 1344–47.

Index

Page numbers in *italics* refer to illustrations.

motivational drives, 200, 204, 208
motivational systems, 17
motor skills, 4–5, 13, 20
multiple personality disorder, 197
muscles, 20, 40, 43, 94, 115, 136
  facial, 59, 91, 126, 215
music, 8, 32–33, 88, 95, 130, 146, 149,
    194
  performing of, 65–66, 70, 95,
    110–11, 208, 210, 218, 219, 230,
    297n
  teaching of, 194, 210, 222, 228
myelin, 42, 85, 148, 290n

Nabokov, Vladimir, 233
narcissism, 210, 216
narrative, 50, 70, 113
  coherent, 70, 172, 175, 182, 188,
    295n
  dismissing, 217, 219–20
  "earned secure," 175, 188
  integration of, 73–74, 268
  live story, 48, 50, 73–74, 102–6, 165,
    167, 171–73, 189, 232
  preoccupied, 217, 225
natural disasters, 240–41, 245, 253
nature, 8, 255
nervous system, viii, 8, 22, 27, 38, 43,
    54–55, 72, 75, 84–85, 267,
    268–69
  interaction of immune system
    with, 44, 55
  see also autonomic nervous system
neurobiology, 51–52
  see also interpersonal neurobiology
neurochemicals, 42, 44, 82, 133, 157,
    192
neurogenesis, 5, 41–42, 85, 110, 111,
    133, 148
neuromodulators, 133
neurons:
  axonal lengths of, 40, 42, 290n
  creation of, 5, 41–42, 85, 110, 111,
    133, 148
  synaptic linkage of, 38–44, 60, 75,
    85, 110, 117–19, 133–34, 139, 150,
    290n
  see also brain, neural circuits of;
    mirror neuron system

neuroplasticity, 5, 38–44, 70, 83,
    84–85, 87, 104, 110, 111,
    133–34, 147–49, 245–47, 269,
    273n, 274n, 275n, 284n, 285n
neuroscience, vii, xiv–xv, xvii, xviii,
    38, 39, 51, 55–57, 59, 83, 85–86,
    107, 149, 152, 274n, 276n,
    277n
neurotoxins, 81
neurotransmitters, 28, 40, 82, 83,
    148
nightmares, xi, 158–59, 293n
noradrenaline, 44, 82
novelty, 40, 41, 84–85, 113
nucleus basalis, 133, 289n

objectivity, 31–36, 95–97, 269
observation, 31–36, 94–95, 99, 269
obsessive-compulsive disorder
    (OCD), 75, 84, 241–51, 298n
  possible genetic link to, 243, 244
  treatment of, 241–51, 298n
occipital lobe, 20, 21
occupational therapy, 51
O'Donohue, John, 71, 282n
oncology, 48
openness, 41, 62, 64, 92, 130, 139,
    188, 206, 269
  bodily, 63, 73, 118, 225–26, 229
  emotional, 111, 228–31
  reflection and, 31–36, 97–98,
    269
  in relationships, 75, 188
optimism, 132
orbitofrontal cortex, 16, 22, 269,
    276n–77n, 280n
orientation, 103
orthopedics, 159
oxygen, 38
oxytocin, 44, 278n

pain, xiii, 39, 121
  emotional, 6–7, 10, 12–13, 31, 34,
    67, 124–26, 132, 195–97
  escape from, 124–26, 157–58
  medications for, 217
  physical, 6, 125, 157, 164
panic attacks, 134–35, 136, 139–41,
    142

About the Author

DANIEL SIEGEL, M.D., is a clinical professor of psychiatry at the UCLA School of Medicine, co-director of the UCLA Mindful Awareness Research Center, and executive director of the Mindsight Institute. A graduate of Harvard Medical School, he is the author of the internationally acclaimed professional texts *The Mindful Brain* and *The Developing Mind,* and the co-author of *Parenting from the Inside Out.* Dr. Siegel keynotes conferences and presents workshops throughout the world. He lives in Los Angeles with his wife and two children.